Contents

Unit 1 Communication, homeostasis and energy (F214)

Module 1 – Communication and homeostasis

Module 2 – Excretion

Module 3 – Photosynthesis

Module 4 – Respiration

Unit 2 Control, genomes and environment (F215)

Module 1 – Cellular control and variation

Module 2 – Biotechnology and gene technologies

Module 3 – Ecosystems and sustainability

Module 4 – Responding to the environment

Introduction

How to use this revision guide

This revision guide is for the OCR Biology A2 course. Here is a plan of the units and modules you will study.

Unit	Module	Length of exam	Total number of marks in the exam (% of total)*
F214: Communication, homeostasis and energy	**Module 1: Communication and homeostasis**	1 hour	60 (15%)
	Module 2: Excretion		
	Module 3: Photosynthesis		
	Module 4: Respiration		
F215: Control, genomes and environment	**Module 1: Cellular control and variation**	1 hour 45 minutes	100 (25%)
	Module 2: Biotechnology and gene technologies		
	Module 3: Ecosystems and sustainability		
	Module 4: Responding to the environment		
F216: Practical skills in Biology 2	Coursework tasks that will be set by OCR and marked by your teachers		40 (10%)

*Remember, 50% of the total marks for Advanced GCE come from the AS course.

When you are revising, have available a copy of the specification. It is written as learning outcomes that tell you exactly what you should learn. As you revise, tick off each learning outcome when you understand what it means, and can describe and explain it.

This book follows the sequence of learning outcomes in the specification. Each module is divided into a number of double-page spreads, each of which ends with some **quick check questions** to test your recall and understanding. At the end of each unit, there are questions that resemble the types of question you will find in your **examination** papers. **Answers** to all questions are on pages 104 to 113.

It is important that you know the meanings of all the terms in the learning outcomes. All these terms, and some others, are highlighted in **bold** in the text.

The AS Biology Units are recommended prior knowledge for the A2 Units. Some of the tips will remind you to revise material from your AS course.

As part of your revision, you may wish to follow up some of the topics in the course. One good source of information – among many books and websites available – is www.askabiologist.org.uk.

We hope you enjoy your course and find this book useful. Good luck with your revision!

Unit 1 (F214) – Communication, homeostasis and energy

This unit of four modules is your introduction to A2 Biology. It is shorter than F215, and you may take it in the January examination session. It contributes 15% to your overall A level qualification. Some of the topics will be familiar to you from GCSE and AS, but some are new, and all are covered in more detail than you experienced at GCSE or AS. Remember that practical work is an important part of your course. Your practical work is assessed in a different unit (F216), but you will need the information given in these pages to analyse the results of your laboratory work.

OCR Revise A2 Biology

Exclusively endorsed by OCR for GCE Biology

Second Edition

Jennifer Gregory, Ianto Stevens
and Richard Fosbery
Series editor: Sue Hocking

www.heinemann.co.uk

✓ Free online support
✓ Useful weblinks
✓ 24 hour online ordering

01865 888080

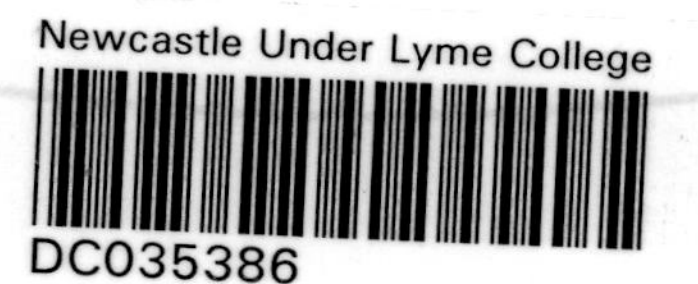

In Exclusive Partnership

Heinemann is an imprint of Pearson Education Limited, a company incorporated in England and Wales, having its registered office at Edinburgh Gate, Harlow, Essex, CM20 2JE. Registered company number: 872828

www.heinemann.co.uk

Heinemann is a registered trademark of Pearson Education Limited

First published 2001
This edition 2008

12 11 10 09 08
10 9 8 7 6 5 4 3 2 1

British Library Cataloguing in Publication Data
A catalogue record for this book is available from the British Library.

ISBN 978 0 435583 73 6

Edited by Anne Sweetmore
Index compiled by Wendy Simpson
Designed by Wearset Ltd, Boldon, Tyne and Wear
Project managed and typeset by Wearset Ltd, Boldon, Tyne and Wear
Original illustrations © Pearson Education Ltd 2001, 2008
Illustrated by Wearset Ltd, Boldon, Tyne and Wear
Cover photo of a false-colour cross-sectional magnetic resonance imaging scan of a human brain
© Neil Borden/Science Photo Library
Printed in China (CTPS/01)

Unit 2 (F215) – Control, genomes and environment

This unit, again of four modules, is longer than F214 and is designed to be taken in the June examination session. It contributes 25% to your overall A2 qualification. As with Unit F214, the information given here may be needed for analysing the results of your practical work for F216.

Topic		Pages in this book	Learning outcomes from the specification	Useful AS revision
Unit 1 (F214) – Communication, homeostasis and energy				
Module 1	Communication	2–3	4.1.1 (a–f)	Unit 1, Module 1: Cell signalling
	Nerves	4–9	4.1.2 (a–k)	
	Hormones	10–13	4.1.3 (a–i)	
Module 2	Excretion	14–17	4.2.1 (a–k)	Unit 1, Module 1: Transport across membranes Unit 1, Module 2: Transport in animals
Module 3	Photosynthesis	18–25	4.3.1 (a–q)	
Module 4	Respiration	26–33	4.4.1 (a–w)	
	End-of-unit questions	34–41		
Unit 2 (F215) – Control, genomes and environment				
Module 1	Cellular control	42–47	5.1.1 (a–j)	Unit 1, Module 1: Cell division and differentiation Unit 2, Module 3: Biodiversity Unit 2, Module 3: Classification and taxonomy Unit 2, Module 3: Evolution
	Meiosis and variation	48–61	5.1.2 (a–s)	
Module 2	Cloning in plants and animals	62–63	5.2.1 (a–f)	Unit 2, Module 1: Nucleic acids Unit 2, Module 2: Enzymes
	Biotechnology	64–65	5.2.2 (a–i)	
	Genomes and gene technologies	66–71	5.2.3 (a–s)	
Module 3	Ecosystems	72–75	5.3.1 (a–l)	Unit 2, Module 2: Food production Unit 2, Module 3: Maintaining biodiversity – conservation
	Populations and sustainability	76–81	5.3.2 (a–i)	
Module 4	Plant responses	82–83	5.4.1 (a–f)	Unit 1, Module 1: Cell signalling
	Animal responses	84–89	5.4.2 (a–l)	
	Animal behaviour	90–93	5.4.3 (a–f)	
	End-of-unit questions	94–103		

Communication and homeostasis

Key words

- receptor
- effector
- cell signalling
- neurone
- hormone
- homeostasis
- negative and positive feedback
- reference (set) point
- ectotherm
- endotherm

Examiner tip

Revise cell signalling from Unit 1 of your AS course.

✓*Quick check 1, 2 and 3*

Communication

All living organisms need to respond adaptively to their environment. In a multicellular organism, different parts of the organism perform different functions. It is therefore essential that an internal communication system controls the activities of these different parts. A communication system between **receptors** and **effectors**, which may be far apart from one another, allows the organism to:

- monitor changes in the internal and external environments
- respond adaptively to such changes
- coordinate the activities of different organs.

Cells communicate with one another through a process called **cell signalling**. The signals may be:

- electrical, by means of nerve cells or **neurones** (see page 4)
- chemical, by means of messengers called **hormones** (see page 10).

Homeostasis

Homeostasis is the maintenance of a relatively stable internal environment within an organism. This provides relatively constant conditions for the individual cells, even when the external environment fluctuates or the organism changes its behaviour, for example by moving more quickly, or by feeding.

Homeostasis requires:

- a **receptor** (sensor or detector) that receives information (input)
- a control mechanism that responds to information and stimulates an effector
- an **effector** to perform the appropriate action.

input → receptor → control mechanism → effector → output

↑———————— **feedback** ←————————┘

Negative feedback occurs when a change in a system sets in motion a sequence of events that counteracts the change and restores the original state. For negative feedback to occur, there must be an optimal norm **reference point** or **set point**. For example, a rise in mammalian blood temperature sets in motion mechanisms to reduce blood temperature, but when the temperature falls below the norm, a new sequence of events raises the temperature. Negative feedback is also seen in mammals in the homeostasis of:

- blood glucose concentration
- ion concentrations of blood
- water content of body fluids
- carbon dioxide concentration of blood.

Positive feedback occurs when a change in a system sets in motion a series of events that result in further change, away from the original state. Positive feedback is rare in biological systems as it can result in an unstable, often extreme, state.

Maintaining a constant core body temperature

Most animals are **ectotherms**. The heat energy that warms them comes from outside their body. Ectotherms may use behavioural means to adjust their temperature. A lizard may move into sunshine to raise its body temperature, or into shade to lower it.

✓*Quick check 4, 5 and 6*

Mammals and birds are **endotherms**, releasing enough heat energy within their body to maintain them above the temperature of their surroundings when necessary. They also maintain a constant body temperature through negative feedback loops, which balance heat loss against heat gain. Like ectotherms, endotherms may use behavioural means such as moving into, or out of, shade to adjust their temperature.

✓Quick check 7

The flow diagram shows the negative feedback loops for temperature control in a mammal.

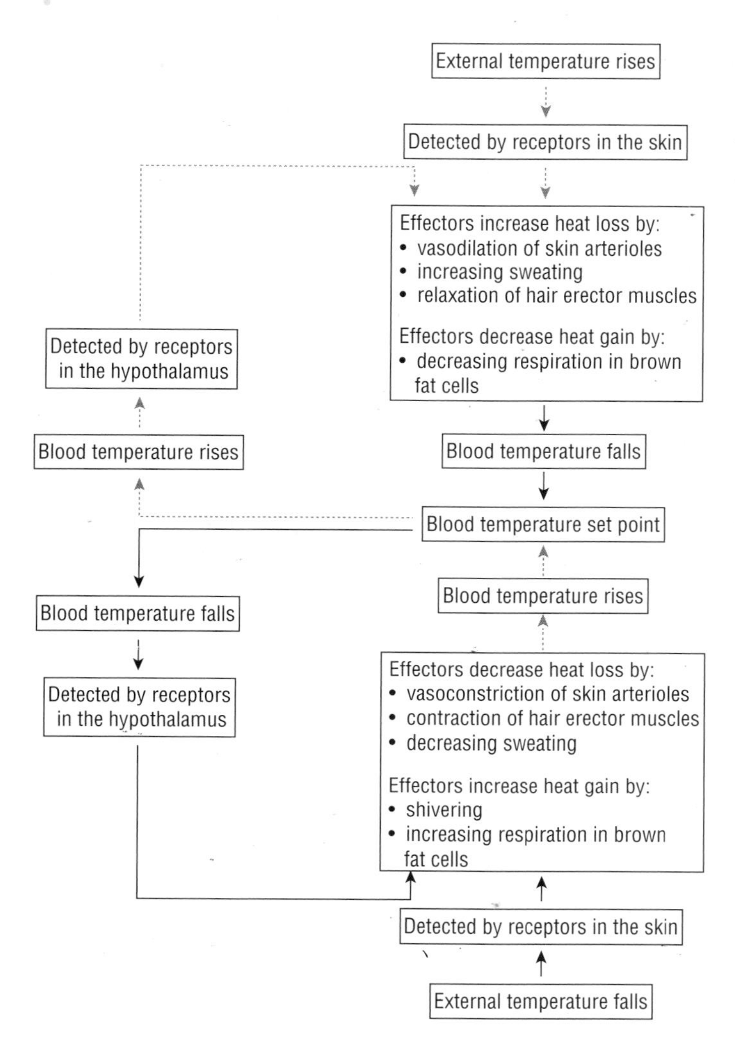

Hint

Above and below certain temperatures this homeostasis fails. Instead, positive feedback results in a high temperature continuing to rise or a low temperature falling still further.

✓Quick check 8 and 9

QUICK CHECK QUESTIONS

1 Explain why organisms need a communication system to respond to environmental changes.
2 State two examples of cell signalling.
3 Explain what is meant by homeostasis.
4 Explain what is meant by negative feedback in a living organism.
5 Distinguish between negative feedback and positive feedback.
6 Explain why positive feedback mechanisms are rare in living organisms.
7 Describe one behavioural response to a falling core temperature in an ectotherm.
8 Name the receptors used by an endotherm to detect a change in core body temperature.
9 Name the effectors used by a mammal to reduce a rising core body temperature.

Nerves

Module 1

Key words

- sensory receptor
- transducer
- sensory neurone
- motor neurone
- dendron (dendrites)
- axon
- myelin
- Schwann cell
- node of Ranvier

Sensory receptors

A **sensory receptor** is a cell in which a change in the internal or external environment produces a nerve impulse or action potential (see page 6). Some receptors are scattered in appropriate body tissues, such as the skin or the lining of the airways. Others are concentrated in organs such as the eye or ear. All receptor cells are **transducers**, in which one form of energy, the stimulus energy, is converted into another form, electrical energy.

Receptors in mammals are adapted to respond to different aspects of the environment, such as:

- light intensity and wavelength
- sound
- touch and pressure
- temperature
- chemicals.

Type of receptor	Type of stimulus energy	Stimulus
Photoreceptor	Electromagnetic	Light intensity and wavelength
Electroreceptor	Electromagnetic	Electricity
Mechanoreceptor	Mechanical	Sound, touch, pressure, gravity
Thermoreceptor	Thermal	Temperature change
Chemoreceptor	Chemical	Humidity, smell, taste, water potential, ion concentration

✓Quick check 1 and 2

Each type of receptor must have the means of absorbing its particular stimulus energy. For example, a rod cell in the retina, which can respond to the presence of light but cannot detect its wavelength, absorbs light by means of a pigment, rhodopsin. (The pigment appears slightly purple in white light and is also called visual purple.) When light energy is absorbed by rhodopsin, the pigment molecule changes shape. This, in turn, leads to a change in permeability of the rod cell membrane. One photon of light falling onto a rod cell is sufficient to trigger a response.

Hint

Compare the changing shape of rhodopsin with shape changes in other molecules you have met, such as enzymes.

Similarly, the three different types of cone cell in the retina have slightly different light-absorbing pigments (opsins), which give them different sensitivities to different wavelengths of light. The three types of cone absorb in the blue, green or red wavelengths of the spectrum, and together give colour vision.

A Pacinian corpuscle is a pressure or vibration receptor in human skin. It consists of a sensory nerve ending enclosed in concentric layers of flattened cells. When pressure is applied to these layers, the permeability of the nerve ending to sodium ions is increased, and when the pressure is great enough, a nerve impulse is set up (see page 6).

✓Quick check 3

Neurones (nerve cells)

Cells that are specialised to transmit the electrical signals of nerve impulses very rapidly from one part of the body to another are called nerve cells or neurones. There are three basic types of neurone in the mammalian nervous system:

- **sensory neurones**, which transmit nerve impulses from a receptor to the central nervous system (brain or spinal cord)
- **motor neurones**, which transmit nerve impulses from the central nervous system to an effector such as a muscle or a gland
- interneurones (intermediate or relay neurones), which transmit nerve impulses between other neurones, for example between a sensory and a motor neurone.

✓Quick check 4

Structure of neurones

A mammalian motor neurone has:

- a cell body or centron, with a large nucleus and large quantities of rough endoplasmic reticulum and Golgi bodies
- many short **dendrites** that carry nerve impulses *towards* the cell body
- a very long process called the **axon** that carries impulses *away from* the cell body to an effector.

A mammalian sensory neurone has:

- long processes on either side of the cell body
- a process, the **dendron**, carrying nerve impulses *towards* the cell body from a receptor
- an axon carrying nerve impulses *away from* the cell body, this time to the central nervous system.

✓Quick check 5 and 6

Interneurones have many shorter processes around their cell bodies, carrying impulses from sensory neurones to motor neurones.

The three types of neurone can be seen in the diagram.

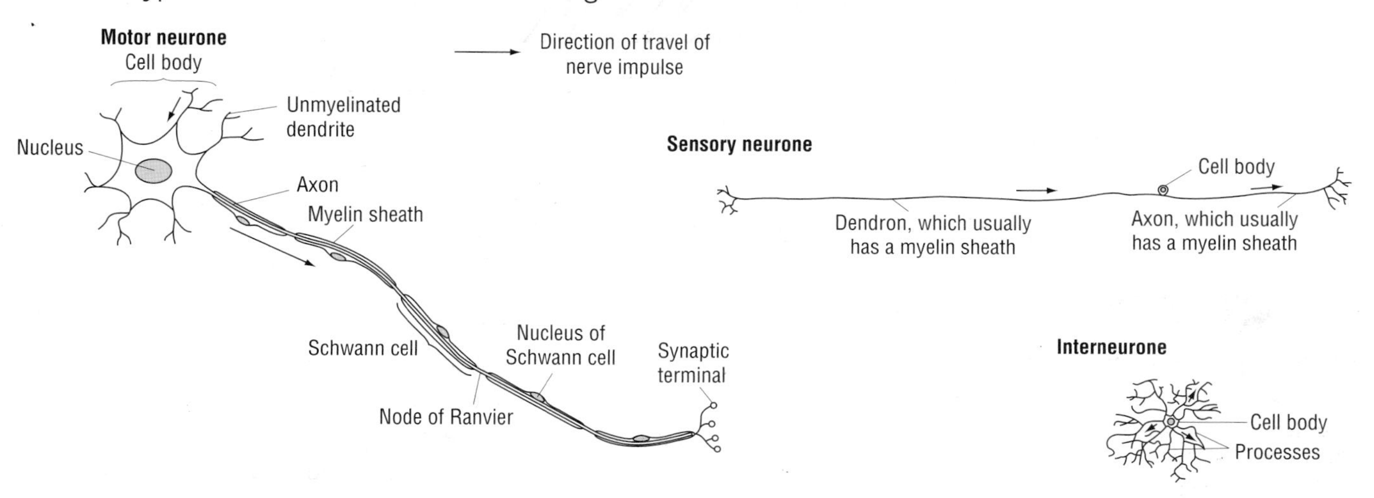

Myelin

Axons and dendrons have cells wrapped around them that nourish and protect them. These cells are **Schwann cells**. In mammals, the Schwann cells wrap themselves many times around an axon or dendron, producing multiple layers of cell surface (plasma) membrane called **myelin**. The gaps between individual Schwann cells are called **nodes of Ranvier**. A myelin sheath helps to speed up the transmission of a nerve impulse from about 4 m s^{-1} to 100 m s^{-1} (see page 7).

✓Quick check 7

QUICK CHECK QUESTIONS

1 Explain why sensory receptors are described as transducers.

2 List the different types of receptor found in mammals.

3 Describe how a named receptor is able to act as a transducer.

4 State the differences in function between a sensory neurone and a motor neurone.

5 Distinguish between an axon and a dendron.

6 List the similarities and differences in structure between a sensory neurone and a motor neurone.

7 State the importance of a myelin sheath to the transmission of nerve impulses.

Transmission of a nerve impulse

Module 1

Key words

- nerve impulse
- resting potential
- action potential
- threshold to fire
- hyperpolarise
- depolarise
- refractory period
- all-or-nothing
- saltatory conduction

Examiner tip

Be careful to talk about a potential difference *across* a membrane.

A **nerve impulse** is an electrical event. Each impulse is a brief change in the potential difference across the plasma membrane of the neurone that passes from one end of the cell to the other. The potential difference across the plasma membrane when the cell is at rest is called the **resting potential**. The brief change in polarity that is a nerve impulse is called the **action potential**.

The resting potential

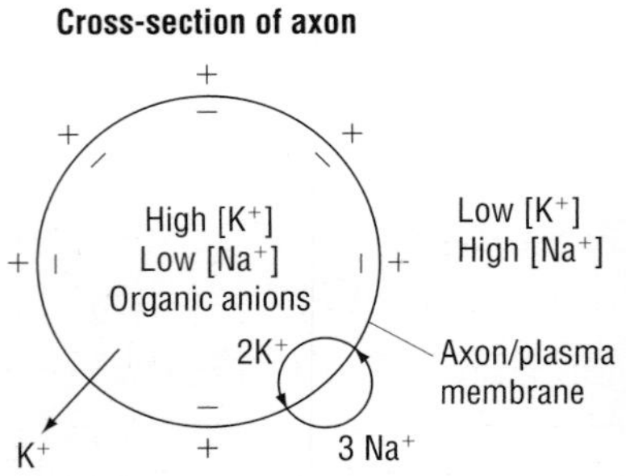

Cross-section of axon

High $[K^+]$
Low $[Na^+]$
Organic anions
$2K^+$
K^+
$3\ Na^+$
Low $[K^+]$
High $[Na^+]$
Axon/plasma membrane

- When *not* conducting a nerve impulse, there is a potential difference (p.d.) across the plasma membrane, such that the inside has a negative charge of –70 mV compared with the outside (see diagram).
- Sodium–potassium pumps in the plasma membrane actively transport K^+ ions into the cell and Na^+ ions out.
- Three Na^+ ions are actively transported out for every two K^+ ions actively transported in.
- The axon contains organic anions to which the membrane is impermeable.
- This, and a slight loss of K^+ ions through the permeable membrane, accounts for the resting potential.
- The membrane is effectively impermeable to Na^+ ions.

✓Quick check 1

The action potential

Cross-section of axon

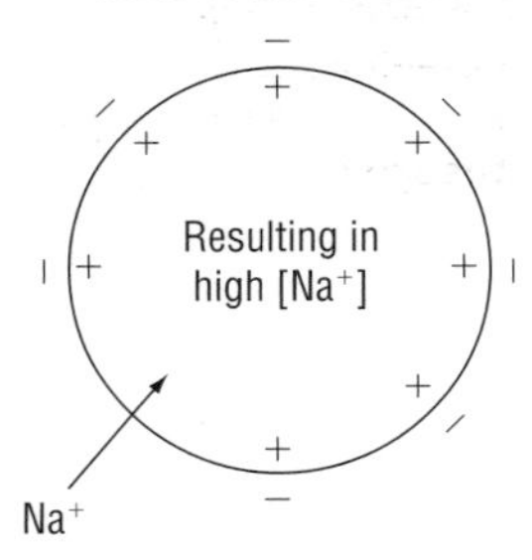

- When the membrane is depolarised and the p.d. across the membrane reaches about –40 mV (the **threshold to fire**), the voltage-gated sodium ion channel proteins in the membrane suddenly open (an example of positive feedback).
- As Na^+ ions are in high concentration outside the cell and in low concentration inside, the Na^+ ions flood into the cell, making it positive inside with respect to outside, and changing the p.d. to +40 mV inside the cell (see diagram).
- The p.d. achieved in a particular neurone is always the same.
- The sodium ion channels now close. Voltage-gated potassium ion channels open and potassium ions diffuse out of the cell. The p.d. becomes negative inside the cell again, falling to about –75 to –90 mV. The membrane is **hyperpolarised**.
- Most of the potassium ion channels close, and the sodium–potassium pump restores the resting potential.
- This sequence of events is an action potential. Transmission of an action potential along an axon forms a nerve impulse.

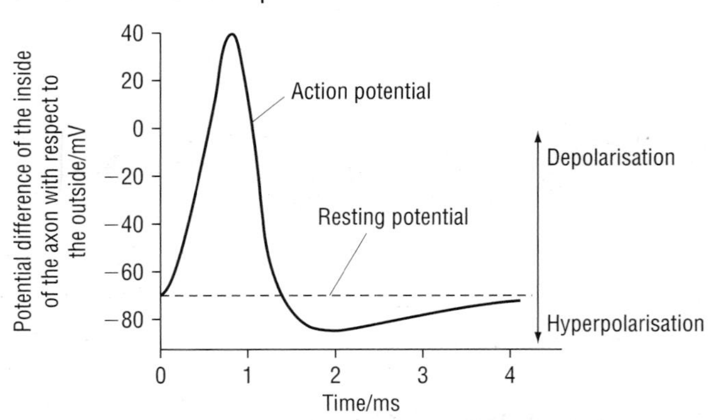

✓Quick check 2 and 3

The changes in potential difference during an action potential are shown in the diagram.

Movement of an action potential

The action potential travels along the plasma membrane. This is caused by the movement of sodium ions, both inside and outside the axon, towards a negatively charged region. This in turn begins to **depolarise** the membrane ahead of the action potential. When the depolarisation reaches the threshold to fire, the sodium ion channels open, giving an action potential.

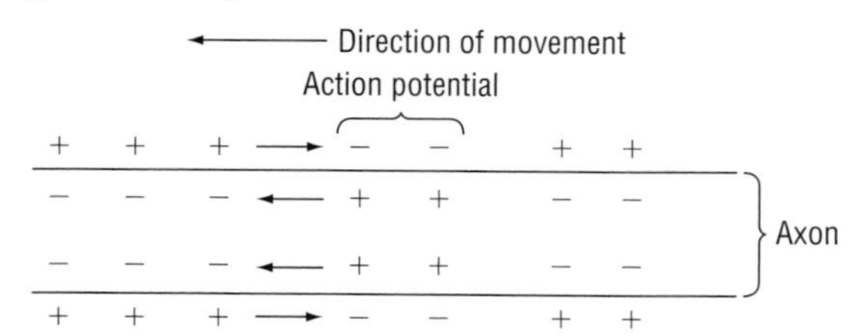

In a non-myelinated mammalian neurone, the action potential is transmitted along the axon membrane at about 4 m s^{-1}; the actual speed depends on the diameter of the axon. A myelinated cell is able to conduct nerve impulses much more quickly by effectively 'jumping' them from one gap (a node of Ranvier) in the insulating myelin to the next (**saltatory conduction**). The nodes are 1–2 mm apart. Local circuits set up by the presence of the action potential at one node depolarise the membrane at the next node to the threshold to fire. The action potential has 'jumped' ahead of the slower wave of depolarisation passing along the axon membrane.

Quick check 4

Examiner tip

Non-myelinated cells have roles in the autonomic nervous system. Myelinated cells are found in the white matter of the central nervous system and in the peripheral nerves of the somatic nervous system (see page 84).

Quick check 5

Refractory period

When the membrane is hyperpolarised, depolarisation will not reach the threshold to fire. The membrane can only respond to depolarisation when its resting potential has been restored. This means there is an unresponsive period: the **refractory period**. This imposes a gap in space and time between one action potential and another, and results in the impulse travelling in one direction.

Examiner tip

Remember that the size of an action potential in a particular cell is always the same.

Information transfer

The *type* of stimulus is recognised by the source of the information. Neurones from particular receptors pass to particular positions in the brain. The *strength* of the stimulus is recognised by the number of cells responding to the stimulus and by the frequency of action potentials on those cells.

When there is sufficient stimulus to depolarise the membrane to the threshold to fire, an action potential follows. When the depolarisation does not reach the threshold to fire, no action potential results. This means that an action potential is **all-or-nothing**. Hence no information about the strength of the stimulus can be transmitted by differences in the size of action potentials. Instead, a strong stimulus results in many action potentials following one another as closely as possible.

Quick check 6 and 7

QUICK CHECK QUESTIONS

1 Explain what gives rise to the resting potential of a neurone.
2 List the events that result in an action potential.
3 Distinguish between depolarisation and hyperpolarisation.
4 Explain how an action potential moves along a non-myelinated axon.
5 Explain how the presence of a myelin sheath speeds up the movement of an action potential.
6 State how different types of stimulus are recognised by the brain.
7 Explain how information about the strength of a stimulus reaches the brain.

Synapses

Key words

- synapse
- neurotransmitter
- acetylcholine (ACh)
- acetylcholinesterase
- cholinergic
- synaptic knob
- synaptic cleft
- synaptic vesicle
- pre- and postsynaptic membrane
- spatial and temporal summation
- facilitation
- depolarise
- hyperpolarise

✓Quick check 1 and 2

A **synapse** is where one neurone makes functional contact with another neurone or with an effector. A neurone may have up to 10 000 synapses with 1000 other cells. Most synapses are between the axon terminals of the stimulating neurone and the dendrites of the stimulated neurone. Typically a gap, the **synaptic cleft**, of about 15 nm is found between the two cells. Most synapses use **neurotransmitters** such as **acetylcholine (ACh)** to cross this gap, but some synapses are purely electrical. Synapses in which the neurotransmitter is ACh are said to be **cholinergic**. The structure of a typical synapse is shown below.

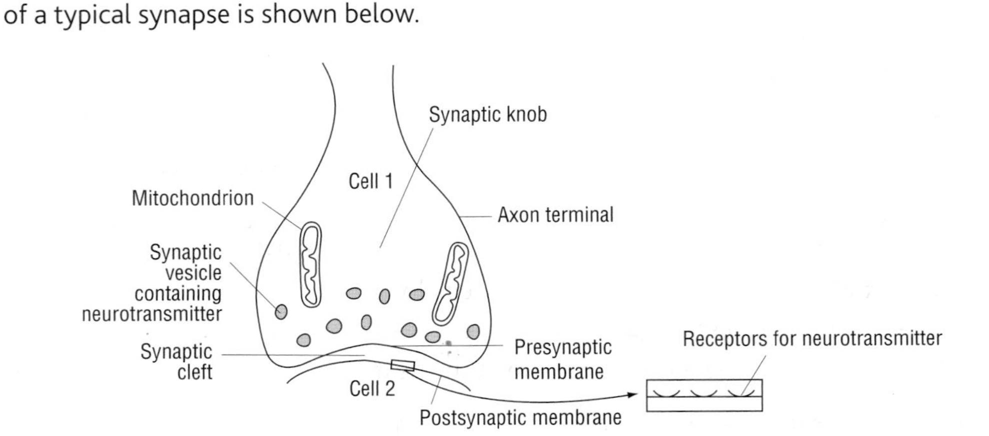

Transmission across a cholinergic synapse

The events taking place during transmission of a nerve impulse across a synapse are summarised below. Relate each numbered event to the figure opposite.

1. An action potential arrives at the axon terminal of cell 1.
2. Calcium ion channels open and calcium ions (Ca^{2+}) diffuse into the axon terminal (**synaptic knob**).
3. This causes the **synaptic vesicles**, which contain ACh made from choline and acetylcoenzyme A (see page 30), to move to the **presynaptic membrane**. This is an active process and needs ATP from the many mitochondria.
4. The vesicles fuse with the presynaptic membrane and release ACh into the synaptic cleft.
5. ACh diffuses across the synaptic cleft and attaches to receptors on the **postsynaptic membrane** (cell 2).
6. This causes the sodium channels to open. Sodium ions enter cell 2, depolarising the membrane and starting a new action potential.
7. The enzyme **acetylcholinesterase** (attached to the postsynaptic membrane) breaks down ACh. Much of the choline is absorbed through the presynaptic membrane and remade into ACh. This too needs ATP.

Examiner tip

pre- = before;
post- = after

Examiner tip

Enzyme names have the ending *-ase*. In this case the enzyme acetylcholinesterase breaks down an organic salt or ester.

✓Quick check 3

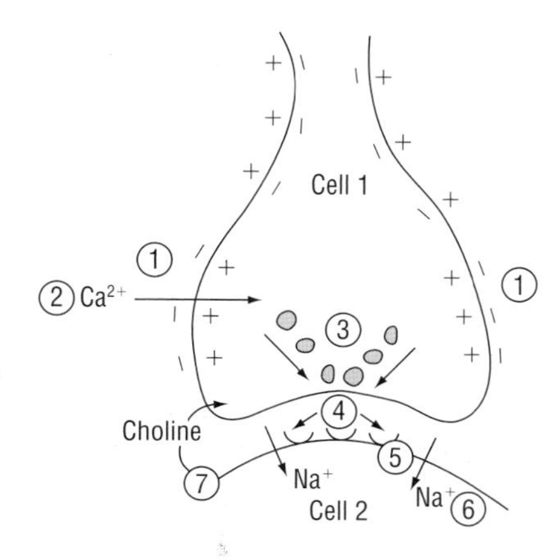

Roles of synapses

Synapses:

- allow transmission in *one* direction only, from presynaptic to postsynaptic membrane
- may be excitatory or inhibitory, providing flexibility of response
- allow **spatial** and **temporal summation** (see below)
- allow **facilitation**, in which the arrival of each action potential leaves the postsynaptic membrane more responsive to the next.

The neurotransmitter of an excitatory synapse **depolarises** the postsynaptic membrane as described above, but that of an inhibitory synapse **hyperpolarises** the postsynaptic membrane, preventing any depolarisation from reaching the threshold to fire (see page 7).

✓Quick check 4

- When two action potentials arrive at two excitatory synapses on the same neurone *at the same time*, the depolarisation of the postsynaptic membrane will be greater. This integration mechanism is spatial summation.
- When action potentials arrive at an excitatory and an inhibitory synapse at the same time, their effects cancel out and no depolarisation of the postsynaptic membrane occurs.
- When two action potentials arrive very closely *after one another*, the depolarising effect is greater. This is temporal summation.

✓Quick check 5 and 6

QUICK CHECK QUESTIONS

1. Explain what is meant by the terms *synapse* and *neurotransmitter*.
2. Draw and label the structure of a typical synapse.
3. List the sequence of events as a nerve impulse is transmitted across a cholinergic synapse.
4. List the roles played by synapses in the nervous system.
5. Explain how a synapse can be inhibitory.
6. Explain what is meant by the terms *summation* and *facilitation* in the context of a synapse.

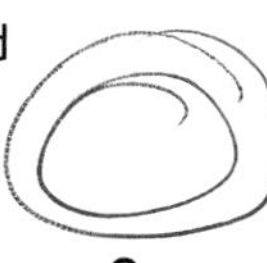

Hormones

Module 1

Key words

- endocrine gland
- exocrine gland
- hormone
- target cell
- target tissue
- first and second messengers
- cyclic AMP
- adrenaline
- insulin
- glucagon
- adenyl cyclase

Quick check 1

Hint

Thyroxine and adrenaline are derivatives of the amino acid tyrosine.

Quick check 2

Examiner tip

Revise cell signalling from your AS course.

Quick check 3

Hint

The presence of a series of enzymes working in sequence in a metabolic pathway means that a single hormone molecule binding to its receptor results in very many phosphorylations and hence a large effect on the cell's metabolism. The effect cascades. Remember the large turnover number of enzymes (see AS Unit 2).

Quick check 4

An **endocrine gland** secretes its product directly into the blood and/or lymph. Together, the endocrine glands form the endocrine system.

An **exocrine gland** has a duct to take the gland's secretion to its site of action, for example salivary glands with ducts to the buccal (mouth) cavity.

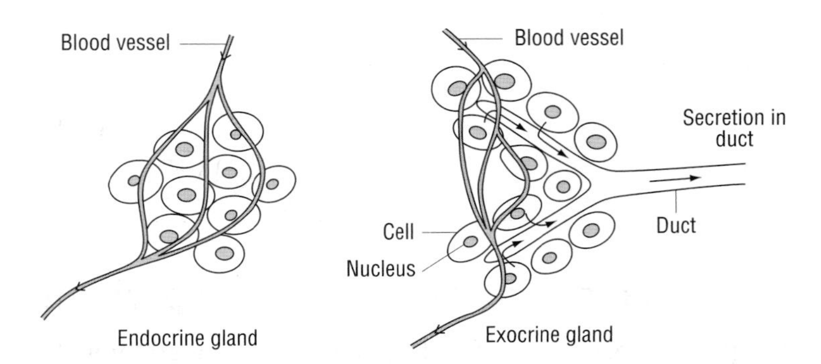

The product of an endocrine gland is a **hormone**. A hormone:

- is a molecule that acts as a chemical messenger
- travels in the blood and lymph all over the body
- causes a specific response in specific **target cells** or **tissues**
- may be a steroid, such as oestrogen, progesterone or testosterone
- may be a protein, such as **insulin** or **glucagon**
- may be derived from an amino acid, like thyroxine and **adrenaline**.

Hormones act in two different ways when they reach a **target tissue**.

- Steroids and thyroxine pass through the plasma membrane and bind with receptors in the nucleus. The hormone–receptor complex binds with DNA and regulates transcription.
- Protein hormones and adrenaline do *not* enter the cells of the target tissue. They bind with receptors on the plasma membrane of the cell. This type of response to the **first messenger** (the hormone) involves **second messengers** within the cell. The second messenger for adrenaline is **cyclic AMP** (cAMP), which is made from ATP by the enzyme **adenyl cyclase**.

When adrenaline reaches a target tissue:

- adrenaline binds to its glycoprotein receptor in the cell surface membrane
- adenyl cyclase is activated, and the concentration of cAMP in the cell increases
- cAMP activates the first of a 'cascade' of enzymes
- the last enzyme in the cascade is a kinase, which adds a phosphate group to an enzyme
- this phosphorylation alters the activity of the enzyme concerned, changing the cell's metabolism
- phosphorylation increases the activity of some enzymes and decreases that of others.

The adrenal glands

The outer (cortex) and inner (medulla) layers of the adrenal glands secrete different hormones with different roles. Some of these are shown in the table.

Endocrine gland	Hormone	Target tissue	Role
Adrenal cortex	Glucocorticoids (e.g. cortisol) Mineralocorticoids (e.g. aldosterone)	Liver Kidney and gut	Stimulates synthesis of glycogen Increases uptake of Na^+ and raises blood pressure
Adrenal medulla	Adrenaline	Heart Liver Smooth muscle of gut	Increases heart rate Stimulates breakdown of glycogen to glucose Inhibits peristalsis

Quick check 5

Hint

Both hormonal and nervous mechanisms are used in the control of heart rate in humans. This involves the hormone adrenaline and also the antagonistic sympathetic (stimulatory) and parasympathetic (inhibitory) nerves to the sinoatrial node (SAN) and atrio-ventricular node (AVN) of the heart. Look up heart structure from Unit 2 of your AS course. Sympathetic and parasympathetic nerves belong to the autonomic nervous system (see page 84).

The endocrine pancreas: the islets of Langerhans

The islets of Langerhans are patches of endocrine tissue scattered throughout the exocrine tissue of the pancreas (see diagram). Remember, the exocrine cells of the pancreas secrete digestive enzymes that travel via a duct to the duodenum. The islets of Langerhans form about 15% of the pancreas.

Each islet has a mixture of several different types of cell, including:

- α cells, which secrete the hormone glucagon
- β cells, which secrete the hormone insulin.

Both α and β cells have receptors in their plasma (cell surface) membranes that enable them to detect changing concentrations of blood glucose.

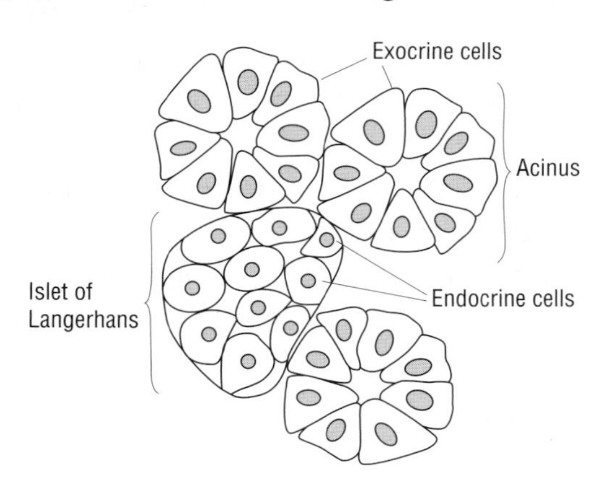

Hint

An islet is a small island or group of cells.

Quick check 6 and 7

QUICK CHECK QUESTIONS

1 Distinguish between endocrine and exocrine glands.
2 Explain what is meant by the term 'hormone'.
3 Describe the two different methods of action of hormones on target tissues.
4 Distinguish between first and second messengers.
5 Describe the functions of the adrenal glands.
6 Explain why the name 'islets' is applied to the endocrine pancreas.
7 Describe the different roles of exocrine and endocrine tissue in the pancreas.

Regulation of blood glucose concentration

Key words

- α and β cells
- diabetes mellitus
- type 1 (insulin-dependent) and type 2 (non-insulin-dependent)

Examiner tip

A healthy human's blood normally contains 80–90 mg per 100 cm^3 (or 800–900 mg dm^{-3}) glucose.

Revise blood plasma from Unit 1 of your AS course.

Blood (plasma) glucose concentration is regulated by negative feedback (see page 2) as shown in the diagram.

β cells secrete insulin
α cells cease to secrete glucagon

Plasma glucose concentration rises
Plasma glucose concentration falls

Plasma glucose concentration of 800 mg dm^{-3}
Plasma glucose concentration of 800 mg dm^{-3}

Plasma glucose concentration falls
Plasma glucose concentration rises

β cells cease to secrete insulin
α cells secrete glucagon

Suppose the plasma glucose concentration *rises* because glucose is absorbed from food in the gut or released from the liver. Then:

- both α and β cells detect the rising concentration of glucose
- α cells cease to secrete glucagon, but β cells secrete insulin
- insulin binds to the receptors in the plasma (cell surface) membranes of muscle, fat and liver cells
- glucose uptake from the blood by these cells is increased
- use of glucose in respiration by the cells is increased
- liver cells convert glucose to glycogen and stop the reverse reaction.

Conversely, when plasma glucose *falls* because it is used rapidly by cells or is lacking in the food eaten:

- both α and β cells detect the falling concentration of glucose
- β cells cease to secrete insulin
- α cells secrete glucagon
- target cells take up less glucose
- the rate of use of glucose decreases and fats or amino acids are used instead in respiration
- glycogen in the liver is converted to glucose and released into the blood.

✓*Quick check 1 and 2*

Insulin secretion

A β cell has a resting potential across its membrane that is kept negative inside the cell by ATP-sensitive potassium ion (K^+) channels. When plasma glucose concentration rises, glucose enters the β cell and is respired, allowing ATP production. The increased ATP concentration in the cell closes the K^+ channels. The membrane depolarises and calcium ions rush into the cell. Insulin is released by exocytosis. In this way, the K^+ channels control the set point for plasma glucose homeostasis (see page 2).

Examiner tip

Compare this mechanism with release of ACh at a synapse (page 8).

✓*Quick check 3*

Diabetes mellitus

Diabetes mellitus arises when no insulin, or not enough insulin, is produced, or when the available insulin does not function correctly.

Insulin-dependent (type 1) diabetes occurs when there is a lack of insulin because of the destruction of all or most of the β cells of the pancreas by an autoimmune response. Regular injections of insulin are needed for survival. The uptake of glucose by brain cells and red blood cells is not dependent on the presence of insulin, but even in the presence of abundant glucose the sugar is not taken up by muscles in the absence of insulin.

Non-insulin-dependent (type 2) diabetes occurs when the body target cells become less responsive to insulin. This type of diabetes can usually be controlled by diet.

✓*Quick check 4*

Human insulin from genetically modified bacteria

There are differences in the amino acid sequence between human insulin and the insulin extracted from cows, sheep and pigs. Most insulin available for injection is from pigs or is human-sequence insulin, either prepared by modifying pig insulin or produced by bacteria into which the gene for human insulin has been added.

Manufacture of human insulin by *Escherichia coli* was one of the early successes in the biosynthesis of human proteins.

The advantages of using human insulin produced by genetically modified bacteria are as follows:

- there is less likelihood of an immune response producing anti-insulin antibodies
- there is a more rapid response, and a shorter duration of response
- patients who have become tolerant of pig insulin (requiring larger doses to have an effect) respond better to human insulin
- some patients do not like the idea of using insulin from animals, for example for religious reasons or because of BSE
- extraction and purification of insulin from animal pancreas is expensive.

Examiner tip

Revise the immune response from Unit 2 of your AS course.

One reported disadvantage of using human insulin is that some patients no longer experience any warning signs of low plasma glucose concentration (hypoglycaemia).

✓*Quick check 5*

Potential use of stem cells to treat diabetes mellitus

In one trial therapy, stem cells were isolated from a patient's bone marrow and the patient was then given chemotherapy to destroy the immune system. Infusion with the stem cells reinstated the immune system. The hope was that the reprogrammed immune system would no longer attack β cells. A problem with this therapy is that insulin is one of the antigens that triggers the immune system to destroy the cells.

In a different approach, human embryonic stem cells (see page 63) have been triggered to produce human insulin in response to rising concentrations of glucose. These cells can be placed in porous capsules to prevent their rejection, and transplanted into the patient's abdomen.

✓*Quick check 6*

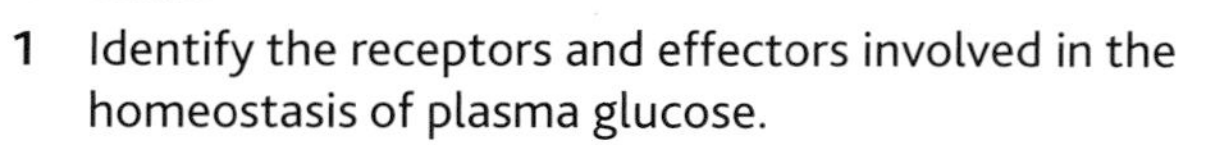

QUICK CHECK QUESTIONS

1. Identify the receptors and effectors involved in the homeostasis of plasma glucose.
2. Describe how plasma glucose concentration is lowered.
3. Describe how a β cell releases insulin.
4. Distinguish between type 1 and type 2 diabetes mellitus.
5. List the advantages of treating diabetics with insulin produced by bacteria.
6. Explain why it is necessary to place stem cells in porous capsules to prevent rejection.

Excretion

Key words

- excretion
- deamination
- ornithine cycle
- urea
- hepatocyte
- lobule
- detoxification
- cortex
- medulla
- nephron
- Bowman's (renal) capsule
- proximal convoluted tubule
- distal convoluted tubule
- loop of Henle
- ascending and descending limbs
- collecting duct
- afferent and efferent arterioles
- glomerulus

Quick check 1

Examiner tip

Revise the structure of amino acids from your AS course.

Quick check 2

Examiner tip

It is important to realise that all hepatocytes can perform all liver functions – there is no division of labour.

Quick check 3 and 4

Excretion is the process by which an organism gets rid of waste metabolic products, such as nitrogenous waste from amino acid metabolism and carbon dioxide from respiration. These waste products are harmful or toxic, and would damage cells if allowed to accumulate.

Carbon dioxide from respiration combines reversibly with water to give carbonic acid, which in turn dissociates to give hydrogen ions and hydrogencarbonate ions:

$$CO_2 + H_2O \rightleftharpoons H_2CO_3 \rightleftharpoons H^+ + HCO_3^-$$

Carbon dioxide is excreted via the lungs. The respiratory centre in the medulla oblongata increases the rate of breathing when the carbon dioxide concentration in the blood rises. Hydrogen ions are removed by the kidneys to maintain blood pH.

Mammals cannot store proteins or amino acids, so any excess is converted into fats or carbohydrates for storage, or for use in respiration. The amino group ($-NH_2$) of each amino acid is removed in the process of **deamination** in the liver, forming the very soluble but very toxic compound ammonia. This is combined with carbon dioxide from respiration, in the **ornithine cycle** in the liver, to give the much less toxic but adequately soluble compound **urea**. This is removed from the blood by the kidneys (see page 16).

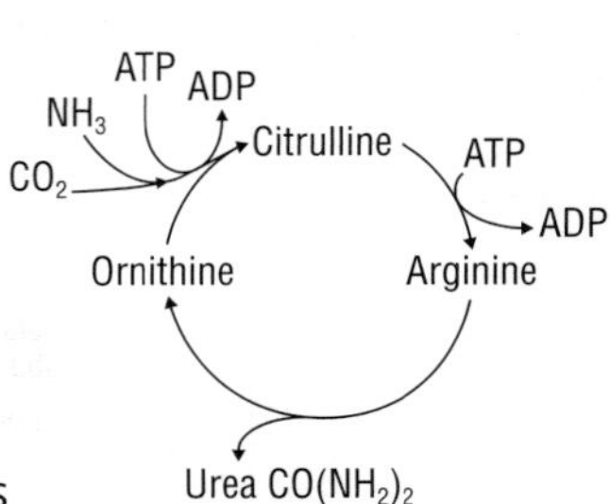

Liver structure

The liver is made up of several lobes. The hepatic artery brings oxygenated blood from the aorta and the hepatic portal vein brings blood from the gut. The hepatic vein takes blood to the vena cava. Liver cells, or **hepatocytes**, are arranged in **lobules** as shown in the diagram.

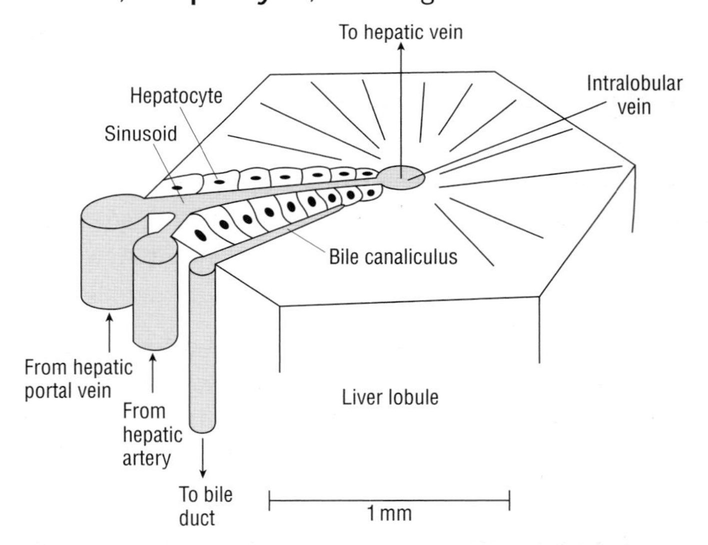

As well as the formation of urea, liver cells perform a large number of different functions, including glucose homeostasis (see page 12). An important function is the **detoxification** of alcohol, antibiotics, steroid hormones and other unwanted or toxic substances. Such substances are taken up by hepatocytes and then broken down. For example, alcohol (ethanol) is absorbed by liver cells. In their cytoplasm, the enzyme ethanol dehydrogenase catalyses the following reaction:

$$CH_3CH_2OH \text{ (ethanol)} \rightarrow CH_3CHO \text{ (ethanal)}$$

NAD ⟶ **reduced NAD**

As reduced NAD accumulates, less NAD is available for other reactions (see page 32). The ethanal is then taken up by the cell's mitochondria, where it is further oxidised to acetate, which may either enter the Krebs cycle (see page 30) or be used to synthesise fatty acids.

Kidney structure

The urinary system is made up of the:

- kidneys
- ureters, taking urine from the kidneys
- bladder, for temporary storage of urine
- urethra, taking urine from the bladder to the outside world.

Each human kidney is made up of around a million tubules, called **nephrons**, and their associated blood vessels. Different parts of the nephrons lie in the outer layer of the kidney, the **cortex**, and in the inner layer, the **medulla**. The kidney receives blood from the aorta via the renal artery and returns it to the vena cava via the renal vein. A longitudinal section (LS) of a kidney and the structure of a nephron are shown below.

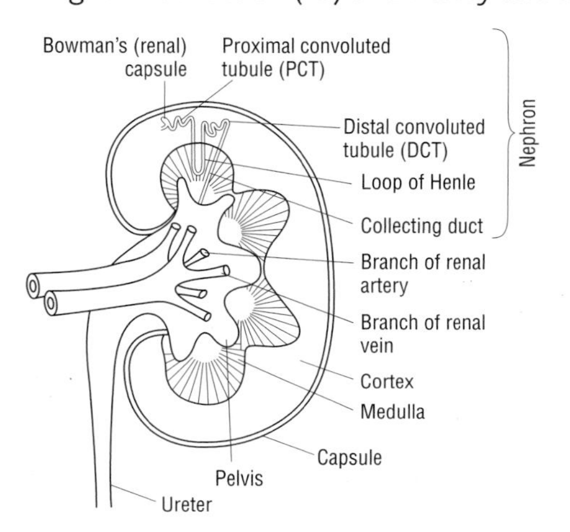

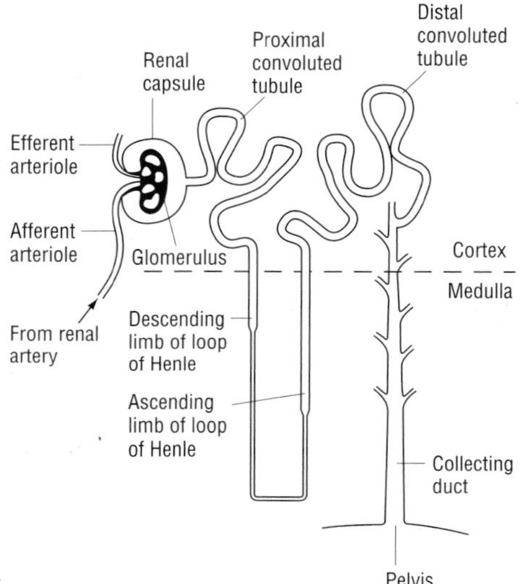

> **Hint**
> *Proximal* means nearer to the starting point; *distal* means further from the starting point.

> **Hint**
> An *afferent* blood vessel carries blood towards a structure; an *efferent* vessel carries it away.

✓Quick check 5 and 6

A nephron begins as a concave **Bowman's capsule** or **renal capsule** in the cortex and leads to a highly coiled **proximal convoluted tubule**. This leads on to a U-tube called the **loop of Henle**, which lies in the medulla. The **ascending limb** of the loop of Henle leads to another coiled tubule in the cortex, the **distal convoluted tubule**, and this leads to a **collecting duct**. A number of nephrons feed into the same collecting duct, which opens into the pelvis of the kidney.

Each Bowman's capsule has an associated **afferent arteriole** bringing blood from the renal artery to a network of capillaries called a **glomerulus**. The **efferent arteriole** draining this network is smaller in diameter than the afferent arteriole. The efferent arteriole divides into capillaries that run close to the convoluted tubules and loop of Henle before joining as venules to take blood back to the renal vein.

✓Quick check 7

QUICK CHECK QUESTIONS

1. Why is excretion necessary in living organisms?
2. Describe the formation of urea in the liver.
3. Draw a transverse section of a liver lobule and label its associated blood vessels.
4. Describe the blood flow through a liver lobule.
5. Draw a nephron and label its different regions.
6. State which parts of a nephron can be found in a kidney medulla.
7. Trace the route of a red blood cell from renal artery to renal vein.

Kidney function

Key words

- osmoregulation
- ultrafiltration
- selective reabsorption
- filtration pressure
- basement membrane
- podocyte
- countercurrent
- antidiuretic hormone (ADH)
- posterior pituitary gland
- osmoregulation
- renal dialysis

The kidney is concerned with both excretion and homeostasis. It:

- excretes urea and other waste solutes as urine
- regulates the water content of the blood (**osmoregulation**)
- regulates the pH of the blood by excreting hydrogen ions if necessary.

These different functions are performed by different regions of the nephron. Hydrogen ions are excreted by most parts of the nephron when the blood pH is too low.

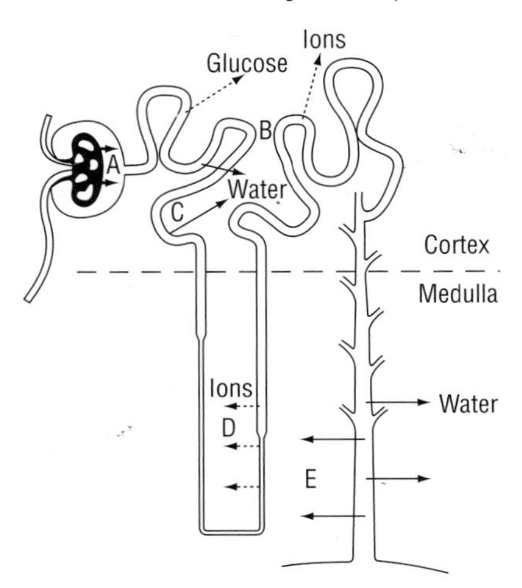

Cortex
A Ultrafiltration from glomerular capillaries into Bowman's capsule.
B Selective reabsorption by proximal convoluted tubule (PCT) and distal convoluted tubule (DCT).
C Absorption of water via osmosis by PCT.

Medulla
D Active transport and diffusion of sodium and chloride ions by loop of Henle to increase solute concentration of tissue fluid outside nephron.
E Reabsorption of water by osmosis in the presence of ADH – osmoregulation.

It is not possible to filter waste material from the plasma selectively. Instead, in the process of **ultrafiltration**, the filtration membrane allows all molecules below a certain size to pass through. To avoid the body being depleted of useful materials, ultrafiltration is followed by a process of **selective reabsorption** of the substances the body cannot afford to lose. At the same time, most of the water lost from the blood during ultrafiltration is returned to the blood, and the hormonally controlled mechanism of osmoregulation takes place.

Ultrafiltration

- The afferent arteriole supplying the capillaries of the glomerulus is wider than the efferent arteriole. This, and the short distance of the kidneys from the aorta, provides the necessary hydrostatic **filtration pressure**.
- The filtration membrane is the **basement membrane** of glycoproteins surrounding the endothelium of the capillaries. The inner layer of the Bowman's capsule consists of highly adapted **podocytes**, which provide support for the membrane with minimum interruption to filtration.
- Water and crystalloid solutes, including urea, are filtered from the blood into the Bowman's capsule. Cells, platelets and colloidal solutes, including the blood proteins with a relative molecular mass greater than 69 000, do *not* pass through the undamaged filtration membrane. Hormones in the blood pass into the filtrate, so urine analysis can test for, say, pregnancy or the misuse of anabolic steroids.

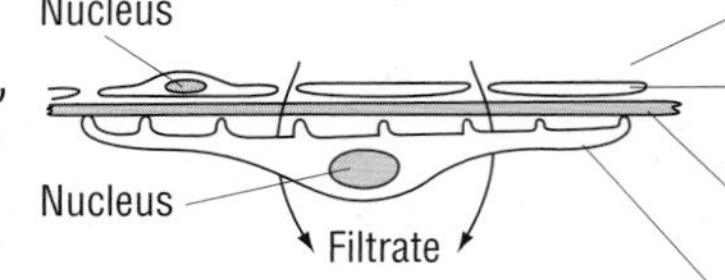

Quick check 1

Hint

Crystalloids are substances that form a true solution. Colloids have particles of about 100–10 000 nm in diameter, which remain dispersed.

Quick check 2

Selective reabsorption

- Glucose, amino acids, vitamins and some ions are reabsorbed by the proximal convoluted tubule (PCT).
- Sodium ions are actively transported out of the PCT cells through the plasma membranes next to blood capillaries, so sodium diffuses into the cells from the filtrate via cotransporter proteins, bringing glucose into the cells with the sodium ions.
- The PCT cells have microvilli to give a large surface area of contact with the filtrate and many mitochondria to provide ATP energy for active transport.
- Ions are absorbed from the filtrate by distal convoluted tubule (DCT) cells, which, although smaller, have the same adaptations as the PCT cells.

Quick check 3

Examiner tip

Remind yourself of the transport of substances in and out of cells from your AS course.

Passive processes

- 65% of the water filtered from the blood by ultrafiltration is reabsorbed at the PCT by osmosis, because the filtrate has a *higher* water potential than the blood plasma it was filtered from.
- Some urea diffuses out of the filtrate at the PCT.
- In the presence of **antidiuretic hormone (ADH)** from the **posterior pituitary gland**, the collecting duct wall is permeable to water, which passes out of the filtrate by osmosis. The release of ADH is triggered by osmoreceptors in the hypothalamus (see page 84). When little ADH is present, the collecting duct wall is impermeable to water. This is **osmoregulation**.

Examiner tip

Review water potential and channel proteins for water (aquaporins) from your AS course.

✔Quick check 4

The countercurrent mechanism

- Sodium and chloride ions move by active transport and diffusion out of the filtrate as it moves along the ascending limb of the loop of Henle.
- This decreases the water potential of the tissue fluid in the tips of the pyramids of the medulla of the kidney, through which the collecting ducts pass.
- The ions pass back into the descending limb and are recycled. This allows them to be concentrated in the medulla.
- The low water potential outside the collecting duct allows water to leave the collecting duct by osmosis, provided ADH is present.

Hint

Fluid flows in opposite directions in the two limbs of the loop of Henle, hence 'countercurrent'.

✔Quick check 5 and 6

Kidney failure

Kidney failure leads to increased retention of water and salts, leading to high blood pressure and swelling (oedema) caused by increased quantities of tissue fluid. If untreated, this can be fatal. It may be treated by **renal dialysis** or by transplanting a kidney.

✔Quick check 7 and 8

In renal dialysis, blood is pumped through partially permeable tubing with pores that allow ions, water and small molecules such as glucose and urea to pass through. The tubing is bathed in dialysis fluid, which has the same concentration of ions and glucose as blood, so that their concentrations in the blood are maintained. The dialysis fluid does not contain urea, so urea diffuses from the blood into the dialysis fluid.

A kidney transplant solves the problem of permanently damaged kidneys, assuming a suitable donor is available. It is essential that the tissue type of the donor kidney closely matches that of the recipient or it will be rejected by the recipient's immune system. Drugs to suppress this rejection (immunosuppressant drugs) are taken.

Examiner tip

You may have used partially permeable tubing such as cellophane or Visking tubing in your AS osmosis experiments. Look up the immune reaction from your AS course.

QUICK CHECK QUESTIONS

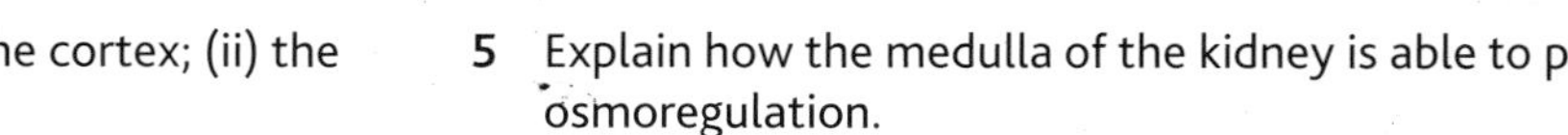

1 List the processes taking place in: (i) the cortex; (ii) the medulla of the kidney.

2 What is the role of a podocyte in ultrafiltration?

3 Explain why the process of selective reabsorption is necessary in the kidney.

4 Explain the return, at the PCT, of much of the water filtered from the blood at the Bowman's capsule.

5 Explain how the medulla of the kidney is able to perform osmoregulation.

6 State the roles of the hypothalamus and posterior pituitary gland in osmoregulation.

7 State two ways of treating kidney failure.

8 Suggest why oedema may be fatal if untreated.

Photosynthesis

Key words

- autotroph
- heterotroph
- light-dependent
- light-independent
- stroma
- thylakoid
- lamella
- rubisco
- granum

Examiner tip

Review autotrophs and heterotrophs from your AS course.

Quick check 1

Autotrophs and heterotrophs

An **autotroph** is an organism that can use an external energy source and simple inorganic molecules to make complex organic molecules. The source of carbon for the organism is carbon dioxide. Autotrophs that use light energy, such as plants, some protoctists and photosynthetic prokaryotes, are called *photoautotrophs*. Some autotrophs use chemical energy sources: these *chemoautotrophs* include the nitrifying bacteria, which obtain their energy by oxidising ammonia to nitrite or nitrite to nitrate.

A **heterotroph** is an organism, e.g. an animal or fungus, that needs to take in complex organic molecules (such as carbohydrates, proteins and lipids) to act as a source of both energy and usable carbon compounds.

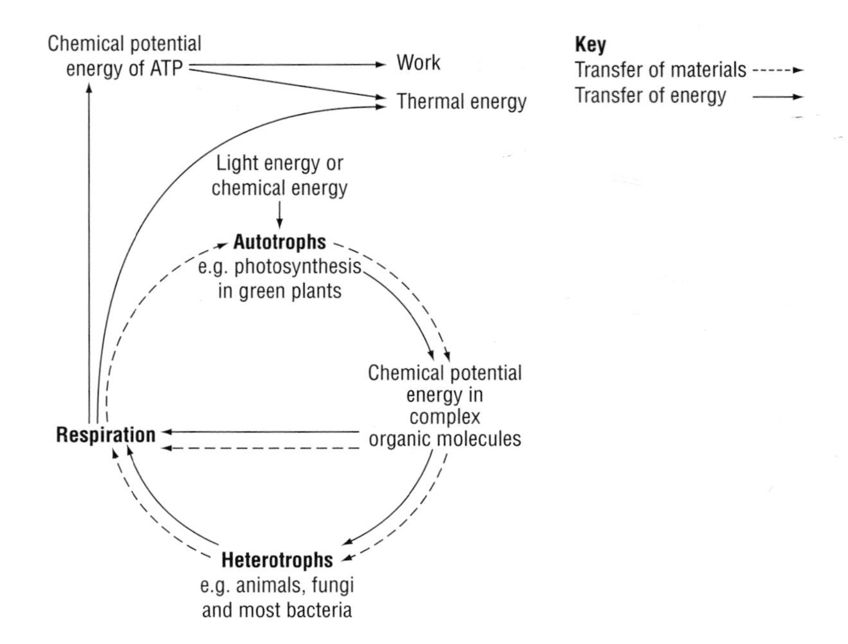

Examiner tip

Follow the one-way transfer of energy first, then the cyclic transfer of materials.

Quick check 2

The relationship between photosynthesis and respiration, and the transfer of energy and materials from autotroph to heterotroph, can be seen in the figure. The energy flows in one direction – it cannot be recycled. Materials such as the elements carbon and nitrogen are recycled.

Hint

Fixation of an element is the conversion of an inorganic source of the element to an organic one.

Photosynthesis

Photosynthesis is the trapping or fixation of carbon dioxide followed by its reduction to carbohydrate using hydrogen from water. The necessary energy comes from absorbed light energy.

$$n\mathrm{CO_2} + n\mathrm{H_2O} \xrightarrow[\textbf{chlorophyll}]{\textbf{light energy}} (\mathrm{CH_2O})n + n\mathrm{O_2}$$

Two sets of reactions are involved:

- **light-dependent** reactions in the chloroplast **lamellae** (**thylakoid** membranes)
- **light-independent** reactions in the chloroplast **stroma**.

Examiner tip

Do not call the light-independent reactions 'dark reactions'. They do not need darkness; they simply do not themselves require light.

Quick check 3 and 4

Chloroplast structure

A chloroplast is the eukaryotic organelle adapted for photosynthesis. In broad-leaved plants (dicotyledons), chloroplasts are biconvex discs about 3–10 μm in diameter. Each has:

- a surrounding envelope of inner and outer phospholipid membranes
- a liquid matrix or stroma, where the Calvin cycle takes place and which contains the enzymes needed for the light-independent reactions, including **rubisco** (see page 22)
- a series of flattened, fluid-filled, membranous sacs or thylakoids, which in places form stacks called **grana** connected by lamellae
- small (70S) ribosomes, as in bacteria
- DNA circles (loops)
- lipid droplets and starch grains.

The lamellae, and especially the grana, provide a very large surface area for holding the light-absorbing photosynthetic pigments (see page 20) and the electron carriers and enzymes needed in the light-dependent reactions of photosynthesis.

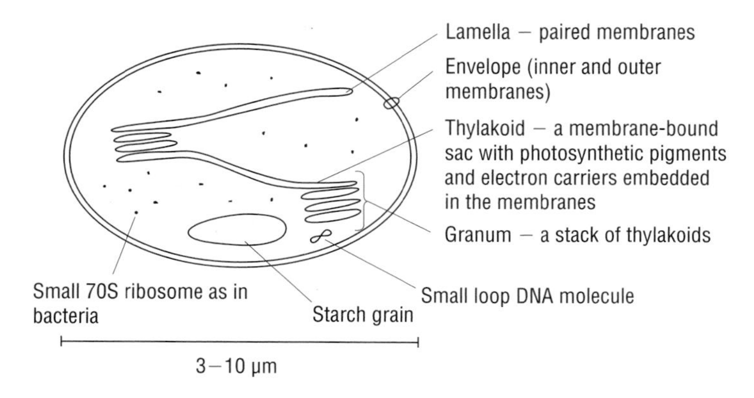

Examiner tip

70S is a measure of size (Svedbergs are a non-SI unit). 70S ribosomes, from prokaryotes, are smaller than 80S, from eukaryotes. You will not be asked about these units.

Quick check 5

Examiner tip

Compare the structure of a chloroplast with that of a prokaryotic cell studied in your AS course.

Quick check 6

Module 3

QUICK CHECK QUESTIONS

1 Explain the different requirements of an autotroph and a heterotroph.

2 Describe the relationship between photosynthesis and respiration in living organisms.

3 Write out an overall chemical equation for photosynthesis.

4 Name the two sets of reactions involved in photosynthesis, and state where in the chloroplast each occurs.

5 Explain why the very large surface area provided by the grana is important in photosynthesis.

6 Draw and label a section of a chloroplast to show its structure.

Photosynthesis: light-dependent reactions

Key words

- photolysis
- NADP
- absorption spectrum
- action spectrum
- photosystem
- cyclic and non-cyclic photophosphorylation
- Z scheme

Examiner tip

NADP is a slightly different form of NAD (see page 28), with a phosphate group attached to one of the ribose sugars.

NADP is a coenzyme. Review coenzymes from Unit 2 of your AS course.

✓Quick check 1

✓Quick check 2 and 3

Examiner tip

Note that the absorption peaks for the chlorophylls are in the violet-blue and red regions of the spectrum.

✓Quick check 4

In the light-dependent reactions of photosynthesis:

- light energy is trapped
- ATP is synthesised in photophosphorylation
- water is split by **photolysis** to give hydrogen ions (H^+) that reduce the hydrogen carrier molecule **NADP** and electrons that pass to chlorophyll (see page 21).

Oxygen is a waste product of photolysis.

Light energy is trapped in the chloroplast lamellae by photosynthetic pigments that are either:

- chlorophylls, e.g. chlorophylls *a* and *b*
- *or* carotenoids, e.g. β-carotene and xanthophyll.

Different pigments absorb different wavelengths of light, and the absorbance by a pigment can be plotted as its **absorption spectrum.** The effectiveness of the different wavelengths of light in promoting photosynthesis can be shown as an **action spectrum**, as shown below.

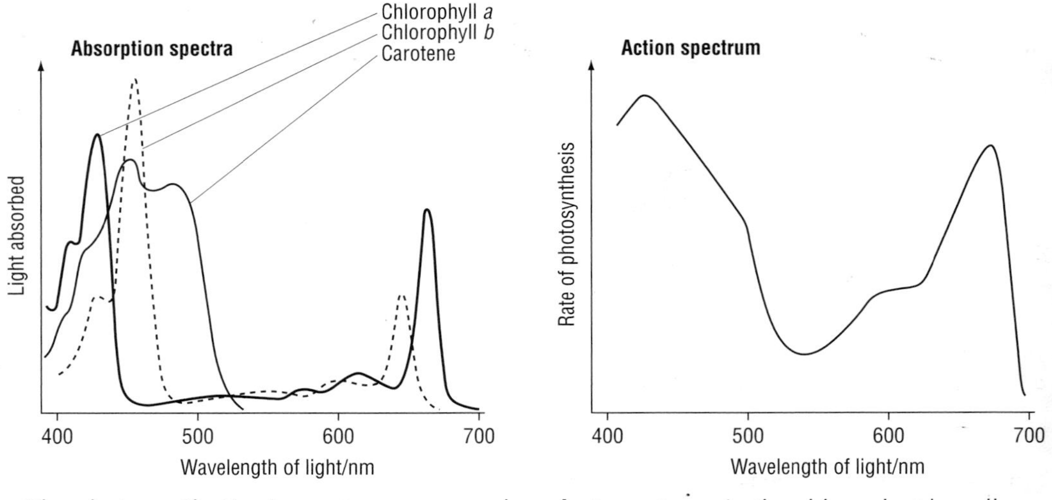

The photosynthetic pigments are arranged as **photosystems** in the chloroplast lamellae.

A photosystem is a light-harvesting cluster of pigments. Each consists of:

- a central *primary pigment* molecule (P), which is one of two forms of chlorophyll *a* with an absorption peak at either 700 nm (P_{700}) in photosystem I or 680 nm (P_{680}) in photosystem II
- surrounding *accessory pigments*, which are other forms of chlorophyll *a*, chlorophyll *b* and the carotenoids. The accessory pigments absorb light and pass energy to the primary pigment. These pigments enable the plant to make use of a range of wavelengths of light.

Cyclic photophosphorylation

When a photon of light is absorbed by photosystem I, an electron in the primary pigment (P_{700}) is excited to a higher energy level and is emitted from the chlorophyll molecule. It is captured by an electron acceptor and passed back to the pigment by a chain of electron acceptors. The energy released allows ATP synthesis from ADP and inorganic phosphate (P_i). This is **cyclic photophosphorylation**.

Non-cyclic photophosphorylation

Both photosystems are needed for **non-cyclic photophosphorylation**. Light is absorbed by both photosystems and electrons are emitted from both primary pigments (P_{700} and P_{680}). Electrons are absorbed by electron acceptors and passed along chains of electron carriers called the **Z scheme**. In the sequence of events:

- P_{700} in photosystem I absorbs electrons emitted by photosystem II
- P_{680} in photosystem II absorbs electrons from the photolysis of water by enzymes in photosystem II

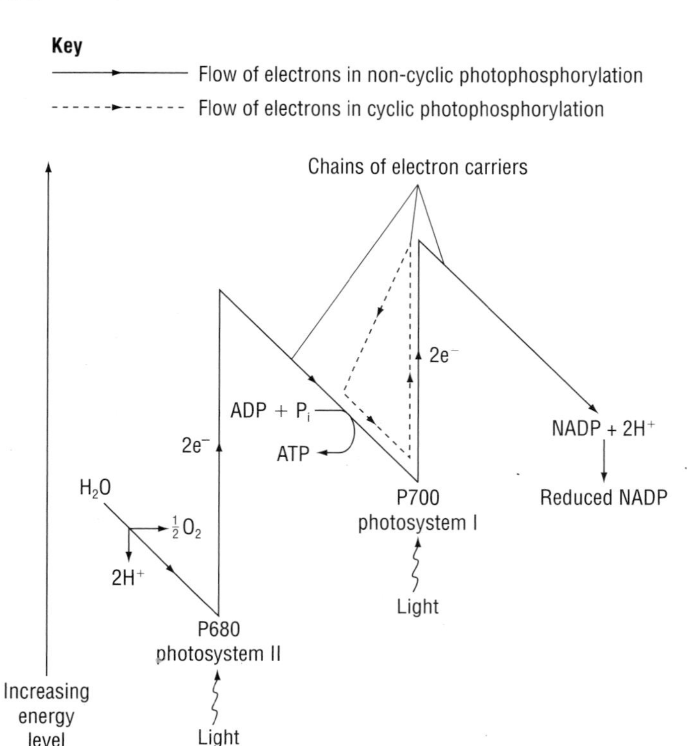

- electrons are used in the synthesis of ATP (see chemiosmosis, page 31)
- oxygen is a waste product.

Comparing cyclic and non-cyclic photophosphorylation:

	Cyclic photophosphorylation	Non-cyclic photophosphorylation
Photosystem	I	I and II
Input(s)	light energy, ADP, P_i	light energy, water, NADP, ADP, P_i
Product(s)	ATP	ATP, reduced NADP, oxygen

Hint

Remember, the Z scheme takes place on the chloroplast lamellae (thylakoids).

✓Quick check 5 and 6

QUICK CHECK QUESTIONS

1. List the products of the light-dependent reaction of photosynthesis.
2. Distinguish between an absorption spectrum and an action spectrum.
3. State the wavelength of visible light at which *least* photosynthesis takes place.
4. State the role of accessory pigments in a photosystem.
5. Describe the different roles of photosystems I and II.
6. Distinguish between cyclic and non-cyclic photophosphorylation.

Photosynthesis: light-independent reactions

Key words

- Calvin cycle
- ribulose bisphosphate (RuBP)
- rubisco
- glycerate 3-phosphate (GP)
- triose phosphate
- hexose

Hint

Rubisco is thought to be the most abundant enzyme in the world.

Module 3

In the light-independent reactions, carbon dioxide is first fixed and then reduced to carbohydrate, using the ATP and reduced NADP produced in the light-dependent reactions. ATP supplies the necessary energy, and reduced NADP supplies the hydrogen atoms needed for reduction. The reactions involved are cyclic: the **Calvin cycle**.

The main steps of the Calvin cycle are shown below.

- Carbon dioxide, which has diffused into the stroma of a chloroplast, is combined with an acceptor molecule that is a 5-carbon sugar, **ribulose bisphosphate (RuBP)**.
- The enzyme needed for this reaction is ribulose bisphosphate carboxylase (**rubisco**).
- An unstable 6-carbon intermediate molecule breaks down into two molecules of the 3-carbon compound **glycerate 3-phosphate (GP)**. Plants in which GP is the first detectable compound to include the newly fixed carbon dioxide are called C_3 plants (because GP is a 3-carbon compound).
- Using ATP from photophosphorylation and hydrogen from reduced NADP, both from the light-dependent reactions, the GP is reduced to **triose phosphate**, a 3-carbon sugar. (Some GP is converted to glycerol or amino acids.) Nitrogen and sulfur are also needed if amino acids, and hence proteins, are to be made; plants take up these elements as inorganic salts – nitrates and sulfates.
- Triose phosphate does not accumulate, but is immediately converted to other products.
- About one-sixth of the triose phosphate is converted to **hexose** (6-carbon sugar), and then to a range of carbohydrates, or is used to synthesise lipids and amino acids.
- Most of the triose phosphate is needed to regenerate RuBP, completing the cycle. This regeneration requires ATP.

✓Quick check 1, 2, 3 and 4

How is RuBP, a 5-carbon sugar, produced from 3-carbon sugars? This process involves a complex cycle including 3-, 4-, 5-, 6- and 7-carbon sugars. The overall result is:

$$\mathbf{5C_3 \rightarrow 3C_5}$$

Of 12 molecules of triose phosphate produced in the Calvin cycle:

- 10 are converted to six molecules of RuBP (the additional phosphate groups needed come from ATP)
- two are *either*
  - converted to one molecule of hexose (which may be used to synthesise starch for storage or cellulose for cell wall construction, or may be converted into other sugars)

  or
  - used to form glycerol, which is converted to lipids.

Examiner tip

Remind yourself of the structures of carbohydrates, fatty acids and amino acids.

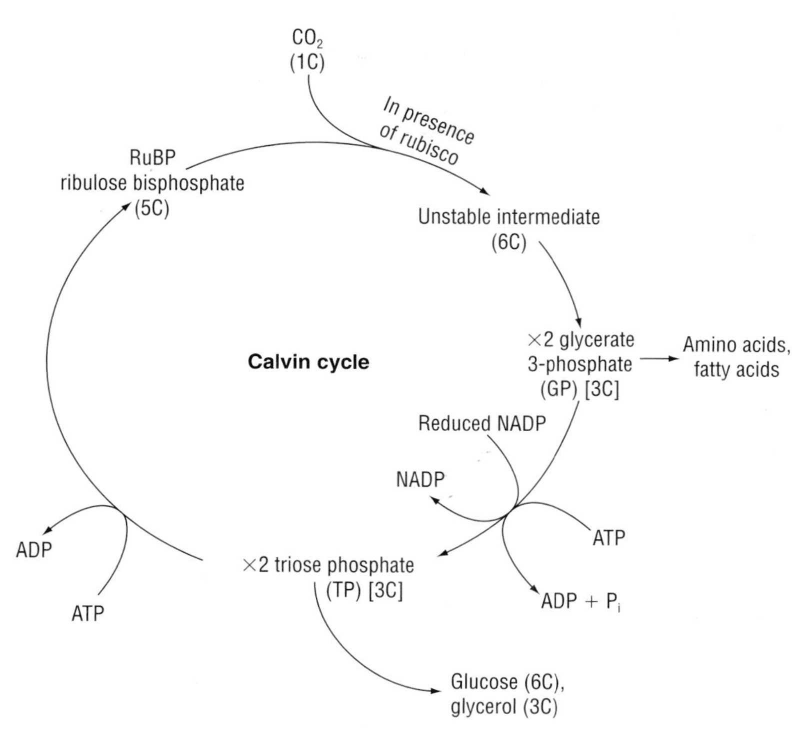

✓Quick check 5 and 6

The steps of the Calvin cycle were determined by first identifying the intermediate compounds and then measuring their relative concentrations in different conditions.

- When the concentration of carbon dioxide (one of the substrates of rubisco) is reduced, then the concentration of RuBP (the other substrate) rises, and the concentrations of GP and triose phosphate fall.
- At low light intensities, lack of ATP and of reduced NADP slows the reduction of GP and so the concentration of GP rises and the concentrations of triose phosphate and RuBP fall.
- Initially, as the temperature rises, all the reactions of the Calvin cycle take place at an increased rate. However, at higher temperatures, rubisco functions as an oxidase. Carbon dioxide and oxygen compete for the active site of rubisco, reducing the amount of carbon dioxide fixed. The concentrations of GP and triose phosphate fall.

QUICK CHECK QUESTIONS

1 What is meant by the term 'carbon dioxide acceptor'?

2 State the reaction catalysed by the enzyme rubisco.

3 State the roles of ATP and reduced NADP in the Calvin cycle.

4 List the possible fates of triose phosphate formed in the Calvin cycle.

5 State what happens to most of the triose phosphate formed in the Calvin cycle.

6 Explain what may happen to the carbon atoms of 12 triose phosphate molecules formed in the Calvin cycle.

Limiting factors

Key words
- limiting factor
- photosynthometer

Quick check 1

When several factors influence the rate of a reaction, the factor that is least favourable will determine the rate at which a reaction can proceed. This is the law of limiting factors – the factor that is most unfavourable is the **limiting factor**.

Limiting factors in photosynthesis

To study the factors or variables that limit photosynthesis, factors must be changed one by one while all other possible variables remain constant. The factors or variables that may limit photosynthesis are:

- light intensity
- concentration of carbon dioxide outside the plant
- temperature (this affects the activity of enzymes catalysing the light-independent reactions and also the rate of diffusion of carbon dioxide)
- degree to which stomata are open or closed
- water supply (which influences stomatal aperture)
- structure of the leaf
- availability of chlorophyll, carrier molecules and enzymes.

Hint

Notice how the line of the graph becomes horizontal at high light intensities.

When the *light intensity* is varied and all other factors remain constant, the rate of photosynthesis increases linearly with increased light intensity over a range of low intensities. The light intensity is the limiting factor. At high light intensities, increasing the intensity has little or no effect (see graph). A factor other than light intensity is limiting the rate of reaction: perhaps the available carbon dioxide.

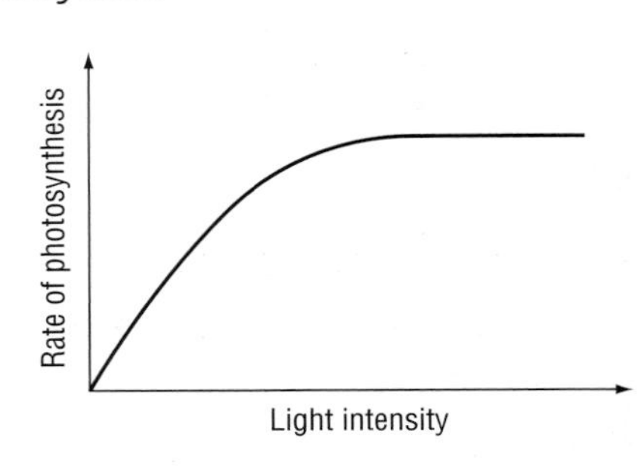

Quick check 2, 3 and 4

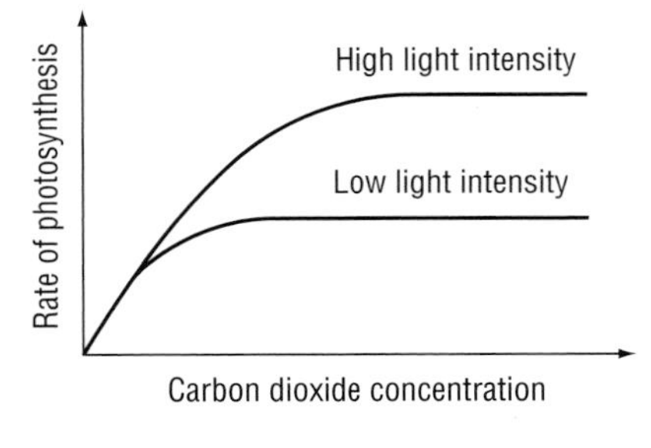

As with increasing the light intensity, increasing the *carbon dioxide concentration* initially increases the rate of photosynthesis. Over this range of concentrations, the carbon dioxide concentration is rate-limiting. At higher concentrations, some other factor is rate-limiting.

Examiner tip

Revise the effect of temperature on enzyme action from Unit 2 of your AS course.

Quick check 5

Quick check 6

At low light intensities, increasing the *temperature* has little effect on the rate of photosynthesis, because light intensity is then the limiting factor. But at high light intensities, increasing the temperature (within limits) increases the rate of photosynthesis, by increasing the rate of the light-independent stages. At higher temperatures (above 25–30 °C) the proteins of the cell begin to be denatured and the rate falls. Below 0 °C, the cells freeze and may be killed by ice crystals. This is why the graph covers a small temperature range.

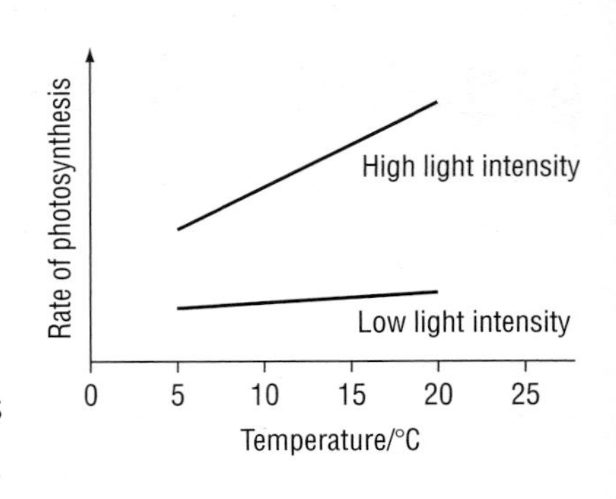

Measuring the rate of photosynthesis

The rate of photosynthesis under different conditions can be measured using a **photosynthometer** (see diagram). By using this apparatus, it is possible to measure the rate of oxygen production by a water plant (e.g. *Elodea canadensis*) while varying the light intensity (or another factor). Bubbles of oxygen collect in the capillary tube of the apparatus. When a suitable volume of gas has been collected in a known time period, it can be drawn alongside the scale by means of the syringe, and the length of the bubble can be measured. This length is directly proportional to volume and can be used instead. Alternatively, if the diameter of the capillary tube is known, the volume of gas can be calculated.

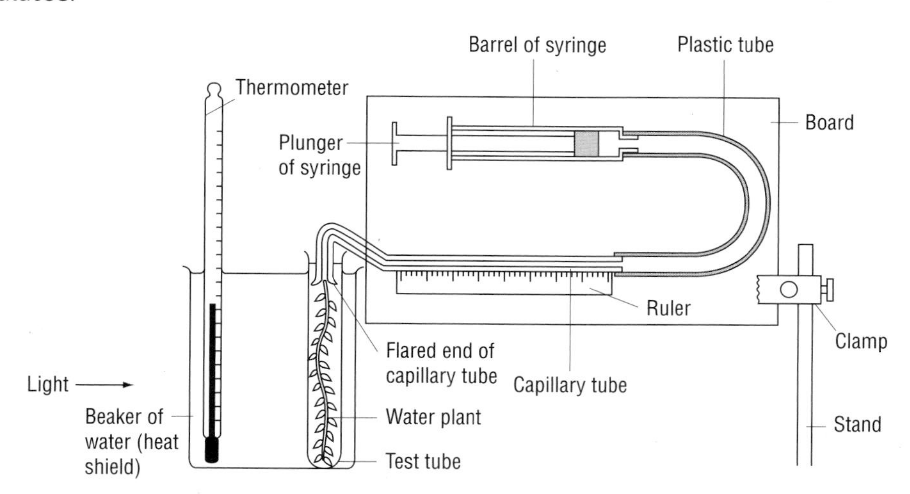

Quick check 7

Note that:

- all variables other than the factor being investigated must be kept constant
- the plant must be left for a time to adjust to new conditions before any measurements are made
- at least three measurements should be taken under each condition, and the mean value found.

QUICK CHECK QUESTIONS

1 Explain what is meant by a limiting factor.

2 List three factors that could limit the rate of photosynthesis at high concentrations of carbon dioxide.

3 Explain why, at low light intensities, the rate of photosynthesis increases with increasing light intensity.

4 Explain why, at high light intensities, increasing the light intensity has little or no effect on the rate of photosynthesis.

5 Copy the graph shown for carbon dioxide and add a curve to show the rate of photosynthesis at an even lower light intensity.

6 Explain why, at low light intensities, temperature has little effect on the rate of photosynthesis.

7 Describe how to investigate experimentally the factors that affect the rate of photosynthesis.

Respiration: ATP

Key words

- metabolism
- anabolism
- catabolism
- work
- adenosine triphosphate (ATP)
- adenosine diphosphate (ADP)
- universal energy currency
- nucleoside
- nucleotide

✓Quick check 1

Module 4

The total of all the biochemical reactions needed for an organism to stay alive is its **metabolism**.

metabolism = anabolism + catabolism

Anabolism is the building up of more complex molecules from simpler ones, for example the synthesis of nucleic acids and carbohydrates. Enzymes are needed for these syntheses of the complex molecules needed for growth. Anabolic reactions are energy-consuming.

Catabolism is the enzymic breakdown of complex molecules to simpler ones. It is the opposite of anabolism. The catabolic reactions of respiration yield energy.

All living organisms need a continuous supply of energy to maintain their metabolism. They must absorb either light energy in photosynthesis or chemical potential energy to do the **work** necessary to stay alive. Such work includes:

- synthesising complex molecules from simpler molecules (anabolic reactions), e.g. polypeptides from amino acids
- active transport of substances across cell membranes against their concentration gradients, e.g. the activity of the sodium–potassium pump
- movement of the whole organism by the action of cilia, undulipodia or muscles, and movement within the organism, e.g. movement of organelles in cells
- maintenance of body temperature, particularly in mammals and birds, which must release thermal energy to maintain the body temperature above that of the environment.

✓Quick check 2

Photosynthesis transfers light energy into chemical potential energy, which can then be released for work by the process of respiration. Both photosynthesis and respiration involve an important intermediary molecule in this energy transfer: **adenosine triphosphate (ATP)**. In many living organisms, most of the energy transferred to ATP is derived originally from light energy; a few prokaryotes (the chemoautotrophs) are not dependent on light energy trapped by photosynthesis but use energy from inorganic chemical reactions instead (see page 18).

ATP

Processes in cells that require energy are linked to chemical reactions that yield energy by an intermediary molecule, **ATP**. Using *one* type of molecule to transfer energy to many *different* energy-requiring processes makes it easier for these processes to be controlled and coordinated. All organisms use ATP as their energy currency: it is a **universal energy currency**.

✓Quick check 3 and 4

ATP is never stored. Glucose and fatty acids are short-term energy stores, while glycogen, starch and triglycerides are long-term stores.

The synthesis and subsequent use of ATP involve energy conversion. The energy ultimately ends up as heat. Some organisms are able to control the rate at which their body gains and loses heat (see page 3).

Hint

An endotherm is an organism that can maintain a constant body temperature, often higher than that of the environment.

Structure of ATP

ATP consists of the organic base adenine and the pentose sugar ribose. Together these make the **nucleoside** adenosine. This is combined with three phosphate groups. ATP is therefore an activated **nucleotide**.

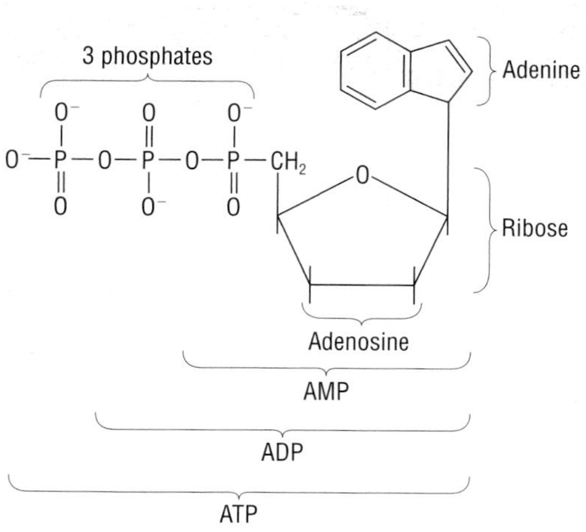

3 phosphates
Adenine
Ribose
Adenosine
AMP
ADP
ATP

> **Examiner tip**
>
> Look up the structures of the nucleotides in DNA and RNA from Unit 2 of your AS course.

✓Quick check 5

The role of ATP

ATP is the standard intermediary molecule between energy-releasing and energy-consuming reactions. Each cell has only a tiny quantity of ATP in it at any one time. The cell does not import ATP, **adenosine diphosphate (ADP)** or adenosine monophosphate (AMP). With few exceptions, each cell must produce its own ATP and recycle it very rapidly. Because it is a small, water-soluble molecule, it is easily moved from where it is made in a cell to where it is needed.

> **Examiner tip**
>
> At one time, the bonds attaching the outer two phosphate groups were called 'high-energy bonds'. This term should not be used because it is misleading. The energy does not come simply from breaking those bonds, but from changes in the chemical potential energy of the whole molecule.

ATP can be synthesised from ADP and an inorganic phosphate group (P_i) using energy, and hydrolysed to ADP and phosphate to release energy. This interconversion is all-important in providing energy for a cell. Hydrolysis of the terminal phosphate group of ATP releases 30.5 kJ mol^{-1} of energy for cellular work:

$$\mathbf{ATP + H_2O \rightleftharpoons ADP + P_i \pm 30.5\ kJ\ mol^{-1}}$$

Removing the second phosphate, giving AMP, also releases 30.5 kJ mol^{-1} of energy, but removing the last phosphate yields only 14.2 kJ mol^{-1}. The energy released comes not simply from these bonds, but from the chemical potential energy of the molecule as a whole.

Module 4

The roles of an ATP molecule include:

- binding to a protein molecule, changing its shape and causing it to fold differently, to produce movement, e.g. muscle contraction
- binding to an enzyme molecule, allowing an energy-requiring reaction to be catalysed
- transferring a phosphate group to an enzyme, making the enzyme active
- transferring a phosphate group to an unreactive substrate molecule so that it can react in a specific way, e.g. in glycolysis and the Calvin cycle
- transferring AMP to an unreactive substrate molecule, producing a reactive intermediate compound, e.g. amino acids before binding to tRNA during protein synthesis
- binding to a trans-membrane protein so that active transport can take place across the membrane.

✓Quick check 6

✓Quick check 7

QUICK CHECK QUESTIONS

1. Distinguish between anabolism and catabolism.
2. What work must a living organism do to stay alive?
3. State the advantage of using ATP as the universal energy currency.
4. Distinguish between an energy currency molecule and an energy storage molecule.
5. State which of the following combinations of molecules gives a nucleotide: (i) ribose + adenine; (ii) ribose + adenine + phosphate.
6. Write a chemical equation for the hydrolysis of ATP to ADP.
7. Identify three different types of work performed by one of your muscle cells.

Respiration: glycolysis

Key words

- glycolysis
- link reaction
- Krebs cycle
- oxidative phosphorylation
- pyruvate
- triose phosphate
- nicotinamide adenine dinucleotide (NAD)
- aerobic respiration

Hint

The material between the plasma membrane and the nucleus of a cell is known as cytoplasm. This includes the various organelles. Cytosol is the matrix in which organelles are suspended.

Module 4

Quick check 1

Hint

Lysis means splitting or breakdown.

Quick check 2

Quick check 3

Hint

The reduction of NAD is represented fully as:

$NAD^+ + 2H \rightarrow NADH + H^+$.

Quick check 4

Respiration is the transfer of chemical potential energy from organic molecules to ATP in living cells. Most cells use carbohydrate, usually glucose, as their fuel. Some cells, such as nerve cells, can *only* use glucose as their respiratory substrate, but others can use fatty acids, glycerol and amino acids.

Respiration of glucose has four main stages:

- **glycolysis** in the cytoplasm (cytosol) of the cell
- the **link reaction** in the matrix of a mitochondrion
- the **Krebs cycle** in the matrix of a mitochondrion
- **oxidative phosphorylation** on the inner mitochondrial membrane.

Glycolysis

The process of respiration begins with the splitting of glucose in the cytoplasm (cytosol) of a cell. After many steps, the 6-carbon (hexose) glucose is converted into two molecules of **pyruvate**, each with three carbon atoms. Energy from ATP is needed in the first two steps, called phosphorylation, but energy that can be used to make ATP is released in the later stages. Glycolysis is summarised below.

- A molecule of glucose is phosphorylated, using two molecules of ATP, to give hexose bisphosphate. This phosphorylation converts an energy-rich but unreactive molecule into one that is much more reactive, the chemical potential energy of which can be trapped more efficiently.
- The hexose bisphosphate is split into two **triose phosphate** molecules.
- Hydrogen atoms and phosphate groups are removed from the triose phosphate, which is oxidised to two molecules of pyruvate (pyruvic acid).

During the process of glycolysis, there is a small yield of ATP by substrate-level phosphorylation (see page 31), giving a *net* gain of two ATP molecules per glucose molecule used. Hydrogens removed during glycolysis are transferred to the hydrogen carrier molecule **nicotinamide adenine dinucleotide (NAD)** (shown below) to give reduced NAD.

NAD is made of two nucleotides linked together. Both nucleotides contain ribose sugar. One nucleotide contains the nitrogenous base adenine; the other has a nicotinamide ring that is the part of the molecule that can accept a hydrogen ion and two electrons. NAD that has accepted these is reduced. NAD is present in cells in small quantities and is continually recycled.

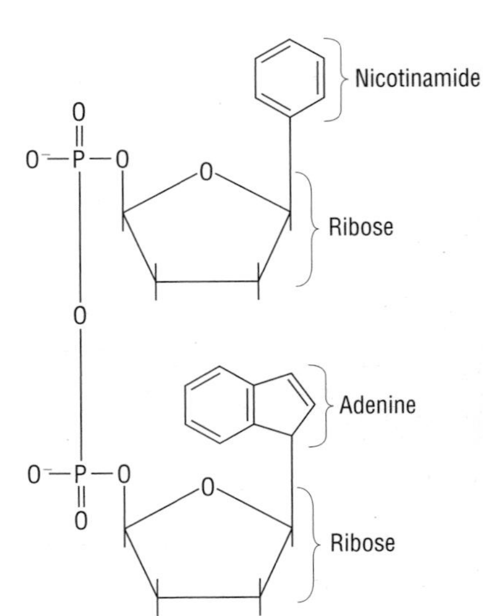

Examiner tip

NAD is a coenzyme. Review coenzymes from Unit 2 of your AS course. You do not need to memorise the structure of NAD.

The main stages of glycolysis are shown in the figure.

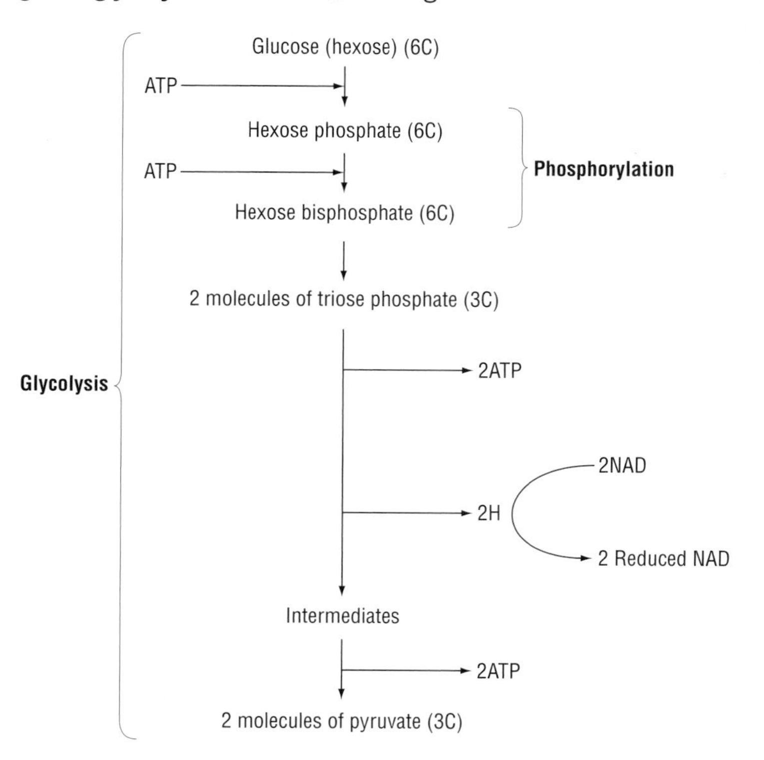

Quick check 5

- Starting point of glycolysis: one molecule of glucose.
- Products of glycolysis: two molecules of pyruvate, a net gain of two molecules of ATP, two molecules of reduced NAD.

Quick check 6

The pyruvate formed in glycolysis is still energy-rich. It passes next to the link reaction. This reaction and all subsequent stages of respiration occur inside a mitochondrion, and can only occur in the presence of free oxygen. Respiration requiring free oxygen is **aerobic respiration**.

Pyruvate is transported into the mitochondrial matrix by a membrane transport protein, which exchanges it for OH^- in the matrix.

Examiner tip

Review transport across membranes from Unit 1 of your AS course.

Quick check 7

QUICK CHECK QUESTIONS

1 State where glycolysis occurs in a living cell.

2 What is meant by a hexose sugar?

3 Explain why phosphorylation of glucose is necessary at the start of respiration.

4 Describe the role of NAD in glycolysis.

5 Draw up a balance sheet of ATP use and production in glycolysis.

6 State the products of glycolysis of one molecule of glucose.

7 What happens to pyruvate when the cell has an adequate supply of free oxygen?

Aerobic respiration

Key words

- acetylcoenzyme A (ACoA)
- flavin adenine dinucleotide (FAD)
- hydrogen carrier
- electron carrier
- electron transport chain
- chemiosmosis
- ATP synthase
- substrate-level phosphorylation

The link reaction

In the link reaction, pyruvate enters the matrix of a mitochondrion and is:

- decarboxylated (carbon dioxide removed)
- dehydrogenated (hydrogen removed)
- combined with coenzyme A to give **acetylcoenzyme A (ACoA)**.

pyruvate (3C) + coenzyme A —(**carbon dioxide**; **NAD** → **reduced NAD**)→ **acetyl (2C) coenzyme A**

Quick check 1

Coenzyme A consists of:

- adenine
- ribose (making a nucleoside together with adenine)
- pantothenic acid (a B vitamin).

Coenzyme A transfers an acetyl group (with two carbon atoms) from pyruvate into the Krebs cycle and plays a central role in respiration. It is present in small quantities in a cell and is recycled.

Hint

NAD and FAD are also coenzymes. NAD and FAD are hydrogen carriers, whereas coenzyme A carries an acetyl group.

Krebs cycle

The Krebs cycle also occurs in the matrix of the mitochondrion. The Krebs cycle is summarised below.

- An acetyl group with two carbon atoms from ACoA is combined with a 4-carbon compound (oxaloacetate) to give a 6-carbon compound (citrate or citric acid). Coenzyme A is re-formed.
- Citrate is then converted back to oxaloacetate in a series of small steps involving decarboxylation and dehydrogenation.

- The carbon dioxide removed is given off as a waste product.
- The hydrogens removed are accepted by NAD or the related **flavin adenine dinucleotide (FAD)**. One FAD and three NAD molecules are reduced during each turn of the cycle. The main role of the Krebs cycle in respiration is to generate a pool of reduced **hydrogen carriers** to pass on to the next stage.
- The regenerated oxaloacetate can combine with another ACoA.
- One molecule of ATP is made directly by **substrate-level phosphorylation** (see page 31) for each ACoA entering the cycle. So two molecules of ATP are made per glucose molecule entering glycolysis.

Examiner tip

The Krebs cycle is also called the citric acid cycle or tricarboxylic acid (TCA) cycle

Quick check 2

Examiner tip

You do *not* need to know the names of the intermediate compounds of the Krebs cycle, only that citrate is formed from acetate and oxaloacetate and then reconverted to oxaloacetate in a series of small steps.

Quick check 3

Electron transport chain and oxidative phosphorylation

These processes take place on and within the inner membrane of the mitochondrion, the cristae.

The energy for phosphorylating ADP to ATP comes from the activity of the **electron transport chain**. Hydrogens from reduced FAD and reduced NAD first pass to hydrogen carriers in the inner membrane and are then split into hydrogen ions (H^+)

and electrons (e^-). The electrons pass along a series of **electron carriers**, each of which is at a lower energy level than its predecessor. The hydrogen ions remain in solution. The final electron acceptor is oxygen. When oxygen accepts an electron, a hydrogen ion is drawn from solution to reduce the oxygen to water. Hence the electron transport chain and oxidative phosphorylation need free oxygen to occur.

✓Quick check 4

The transfer of electrons along the series of electron carriers makes energy available for the synthesis of ATP from ADP and P_i by creating a proton gradient across the inner mitochondrial membrane: this is **chemiosmosis**.

Synthesis of ATP: chemiosmosis

Mitchell's theory of chemiosmosis (1961) is based on the concept that gradients have the power to do work. It requires:

- a membrane across which a gradient can be established – the inner membrane of the mitochondrion or chloroplast
- a gradient – it was known that broken mitochondria showed electron transport, but not ATP synthesis, and Mitchell was able to show that electron flow caused protons to be ejected from inside intact mitochondria
- a device that can use the energy of the gradient to synthesis ATP – stalked particles.

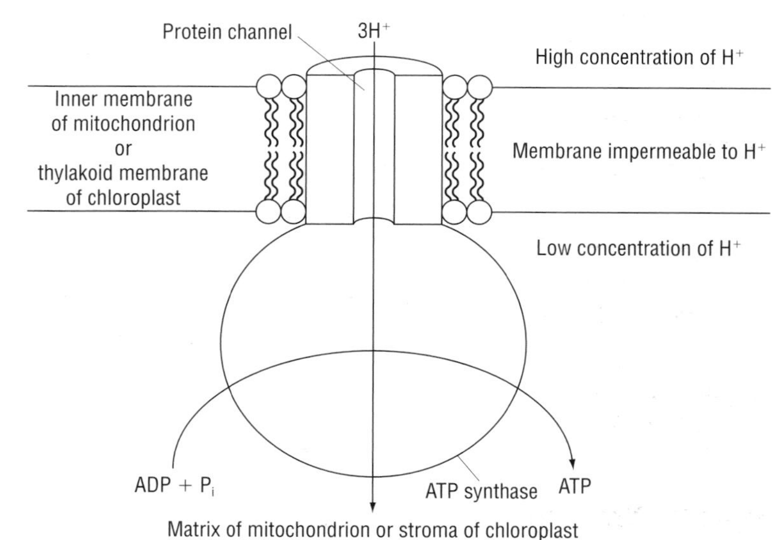

The inner membrane of the mitochondrion is essentially impermeable to hydrogen ions, but the ions flow down their concentration gradient through protein channels spanning the bilayer, as shown in the figure. Part of the protein is the enzyme that synthesises ATP: **ATP synthase**.

Examiner tip

Enzyme names have the ending *-ase*. A synthase synthesises a product, in this case ATP. Note that this process occurs in *both* respiration and photosynthesis.

Potentially, 2.6 molecules of ATP can be made from each reduced NAD entering the electron transport chain, provided that ADP and P_i are available *inside* the mitochondrion. This theoretical yield is rarely achieved because of loss of protons through the mitochondrial membrane and the energy used in transporting materials into the mitochondrion.

✓Quick check 5 and 6

Some ATP is synthesised by a different method. In two stages of respiration, glycolysis and the Krebs cycle, some ATP is synthesised *directly* from the energy released by reorganising chemical bonds. This is called **substrate-level phosphorylation**.

QUICK CHECK QUESTIONS

1 Summarise the events of the link reaction.
2 What is the role of coenzyme A in aerobic respiration?
3 State where each of the following processes occurs in a living cell: (i) link reaction; (ii) Krebs cycle; (iii) oxidative phosphorylation.
4 State the role of oxygen in aerobic respiration.
5 What is meant by chemiosmosis?
6 Describe briefly how ATP is produced using the potential energy of a proton gradient.

Anaerobic respiration and respiratory substrates

Key words

- anaerobic respiration
- alcoholic fermentation
- respiratory substrate

Hint

Lactic acid ($CH_3CHOHCOOH$) occurs in the body as the anion lactate ($CH_3CHOHCOO^-$).

Module 4

When free oxygen is not present, respiration must be **anaerobic respiration**.

In the absence of free oxygen, hydrogen cannot be used up by combining it with oxygen to give water, so reduced NAD cannot be recycled to NAD in this way to allow glycolysis to continue. The stages of respiration inside the mitochondrion cannot occur, but two other pathways (shown below) allow the recycling of reduced NAD formed during glycolysis:

- conversion of pyruvate to ethanol in **alcoholic fermentation**, e.g. by yeast
- conversion of pyruvate to lactate (lactic acid), e.g. by mammalian muscle.

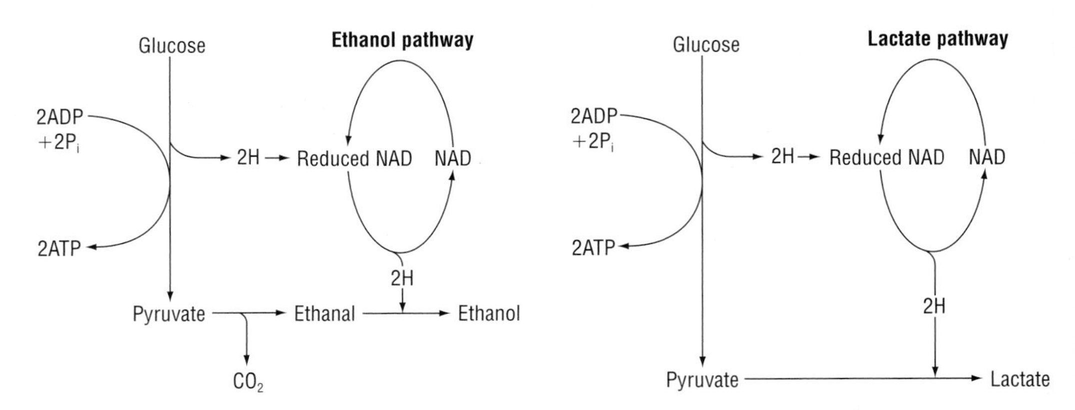

In alcoholic fermentation, pyruvate is decarboxylated to ethanal (CH_3CHO). This accepts hydrogen from reduced NAD and is reduced to ethanol (C_2H_5OH), releasing NAD.

In mammalian muscles that are deprived of oxygen, pyruvate itself acts as the hydrogen acceptor and is converted to lactate. Again, NAD is released.

Quick check 1 and 2

- Both reactions 'buy time' by providing hydrogen acceptors so that NAD is released and glycolyis can continue.
- Both pathways are inefficient and wasteful in that the products (ethanol or lactate) have chemical bond energy that is untapped.
- The ethanol or lactate produced is toxic and restricts the use of the pathways.
- While the lactate pathway is reversible (by the Cori cycle) in the mammalian liver, the ethanol pathway is irreversible.
- There is a net gain of only two ATP molecules per glucose molecule (from glycolysis) during anaerobic respiration.

Quick check 3

The Cori cycle

In mammals, lactate produced during strenuous muscle activity is taken up from blood plasma by the liver, where it is converted to pyruvate and thence to glucose or glycogen. The cycle is shown opposite.

The Cori cycle serves two purposes:

- it 'rescues' lactate and prevents the wasteful loss of some of its chemical bond energy
- it prevents a potentially disastrous fall in plasma pH.

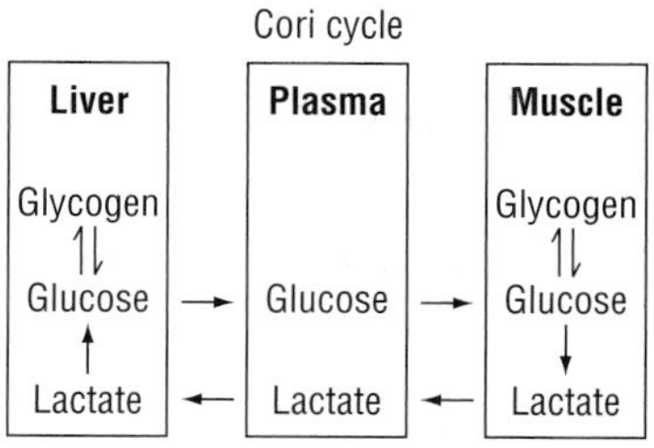

Cori cycle

Liver		Plasma		Muscle
Glycogen ⇅				Glycogen ⇅
Glucose	←	Glucose	←	Glucose ↓
↑ Lactate	←	Lactate	←	Lactate

✓Quick check 4

Respiratory substrates

A **respiratory substrate** is a molecule from which energy can be liberated to produce ATP in a living cell.

Glucose is an essential fuel for some cells, e.g. brain cells, red blood cells and lymphocytes, but some cells, e.g. liver cells, also oxidise lipids and excess amino acids (see opposite). The fatty acid components of lipids are important: carbon atoms are detached in pairs as ACoA and fed into the Krebs cycle (see page 30). Amino acids are deaminated and their carbon–hydrogen skeletons converted into pyruvate or into ACoA. The energy values of these different substrates are not the same.

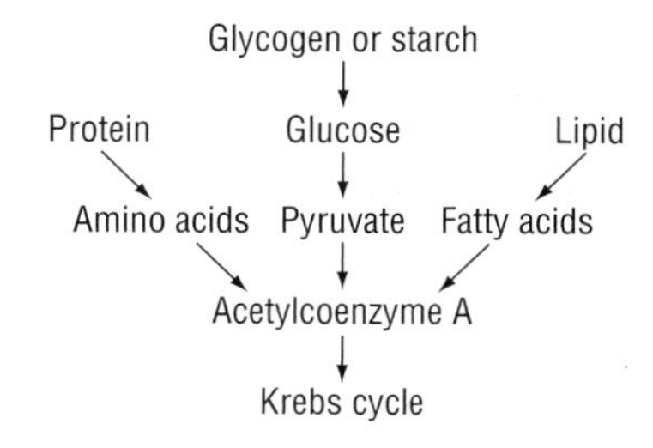

Examiner tip

Look up the structures of fatty acids and amino acids from your AS course.

Module 4

Energy values of respiratory substrates

Most of the energy released in respiration comes from the oxidation of hydrogen to water. The more hydrogens there are in the structure of a molecule, the greater the energy value. Fatty acids have more hydrogens per unit mass than carbohydrates, so lipids have a greater energy value per unit mass. Energy values in kJ g^{-1} are determined by burning a known mass of the substance in oxygen in a bomb calorimeter. Typical energy values are:

✓Quick check 5 and 6

Respiratory substrate	Energy value/kJ g^{-1}
Carbohydrate	16
Lipid	39
Protein	17

✓Quick check 7

QUICK CHECK QUESTIONS

1 Name one cell in which: (i) alcoholic fermentation occurs; (ii) the lactate pathway occurs.
2 Explain how reduced NAD from glycolysis is dehydrogenated in the absence of free oxygen.
3 List the disadvantages of anaerobic respiration in comparison with aerobic respiration.
4 Describe the role of the Cori cycle in mammals.
5 Define the term 'respiratory substrate'.
6 Name two respiratory substrates other than glucose.
7 Explain why lipids have a greater energy value than carbohydrates.

End-of-unit questions

1 Przewalski's horses are found in Mongolia, where the climate is dry and there are great daily and also seasonal variations in temperature. Summer midday temperatures can be higher than 40 °C and winter night temperatures lower than −10 °C. The horses must travel great distances to find grazing and drinking water, which are very scarce. They need to escape from wolves by using intense bursts of speed.

ECG sensors and temperature sensors were implanted under the skin of three adult Przewalski's horses. Heart rate and skin temperature were continuously measured and recorded, using a radio link, without disturbing the animals. Air temperature was also recorded.

Some data from this investigation are shown in the first table. The shaded columns are months during which the mean air temperature was below 0 °C.

	Jan	Feb	Mar	Apr	May	Jun	Jul	Aug	Sep	Oct	Nov	Dec
Mean monthly heart rate/ beats min^{-1}	45	43	45	58	85	80	68	65	68	70	50	43

The researchers concluded that both the physical activity and the basal metabolic rate (BMR) of the horses were lower during the winter.

(a) Explain why heart rate can be a measure of both physical activity (exercise) and BMR. (3)

During the winter, and also in the later part of the night, the skin temperature of the horses often fell to between 28 and 32 °C. The temperature of the internal organs of healthy horses is about 38 °C.

(b) Explain this fall in skin temperature. (2)

In a different investigation, temperatures were recorded from sensors implanted in the body of a horse. The horse showed no ill effects from the keyhole surgery needed to implant the sensors, which were later removed. The records were made several days after implantation.

Some of the data are shown in the second table, when the horse had been standing at rest, and when it had been galloping for 5 minutes.

Site of implantation of temperature sensor	Temperature recorded/°C	
	Horse resting in stable	Horse galloping
Under skin of neck	32.1	36.5
In pulmonary artery	37.8	38.1
In small intestine	37.9	38.0
Between two large hind leg muscles	37.8	40.1

(c) With the help of the table, and using your knowledge of thermoregulation, explain why:

(i) The skin temperature must remain lower than the blood temperature. (1)

(ii) Muscle temperature has increased by about 2 °C in the galloping horse. (2)

(iii) The set point for homeostasis in this horse is just below 38 °C. (1)

Examiner tip

You will need to use ideas from AS unit F211 module 2, and from A2 unit F214 module 1, to answer this question fully.

(d) Describe how the cardiovascular system of a horse would be controlled by nervous and hormonal mechanisms as it responded to the presence of wolves, allowing effective flight. (9)

2 Use the following ideas to help you understand the data on which the parts of this question are based.

- Squid have very large axons, called giant axons.
- The radioactive isotope of sodium is chemically extremely similar to the common, non-radioactive isotope. Radioactive sodium can be used to trace the movement of sodium ions into organisms.
- The radioactivity of a giant axon can be measured with a Geiger counter. The radioactivity in counts min^{-1} is directly proportional to the number of radioactive sodium ions present inside the axon.

A giant axon was placed in a dish, so that the solution in the dish could easily be replaced without stimulating the axon. A solution similar to squid tissue fluid, containing radioactive sodium, was introduced into the dish at hourly intervals. While it was in the radioactive solution, the axon was stimulated using electrodes. The radioactive solution was then replaced by a similar solution containing only non-radioactive sodium. The radioactivity of the axon was measured every 15 minutes until an hour had elapsed. The radioactive solution was then reintroduced and the axon stimulated again, after which the radioactive solution was removed and the cycle repeated. For the first 3 hours, the solutions were rotated hourly as described, but no electrical stimulus was given.

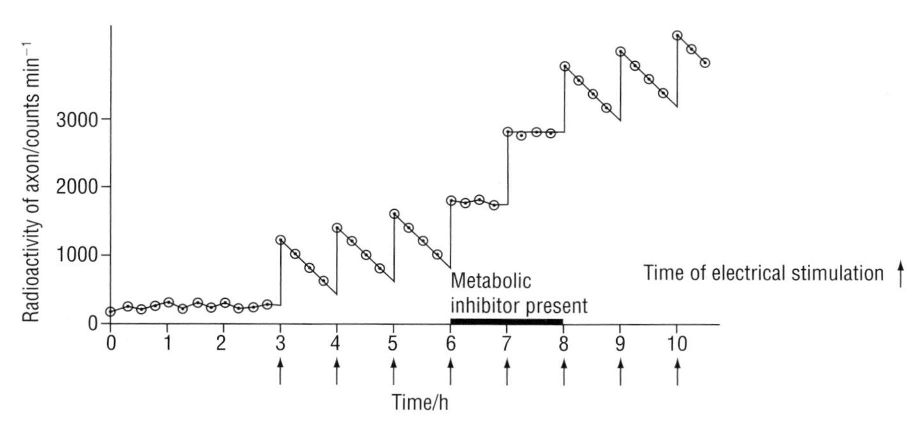

During the period between hours 6 and 8, the giant axon was prevented from synthesising ATP by the use of a metabolic inhibitor. This was added to both radioactive and non-radioactive solutions.

Using the graph, and your knowledge of nerve physiology:

(a) State the effect of an electrical stimulation on the radioactivity of the axon. (2)

(b) Explain why an electrical stimulus produced this effect. (3)

(c) Describe and explain the changes in the radioactivity of the axon between:

(i) 3 and 4 hours after the investigation started

(ii) 7 and 8 hours after the investigation started. (4)

(d) Explain how the results show that an action potential does not require ATP to be present. (1)

Examiner tip

Q1 (d) has 9 marks: you will need to plan an extended answer that uses scientific terms.

Examiner tip

This question is to test understanding and application of knowledge. Don't try to memorise the facts.

Examiner tip

'Using the graph' means giving some figures from the graph – don't forget the units.

The drawing shows a region of nervous tissue in which an axon terminal, A, is in contact with two other nerve endings, B and C.

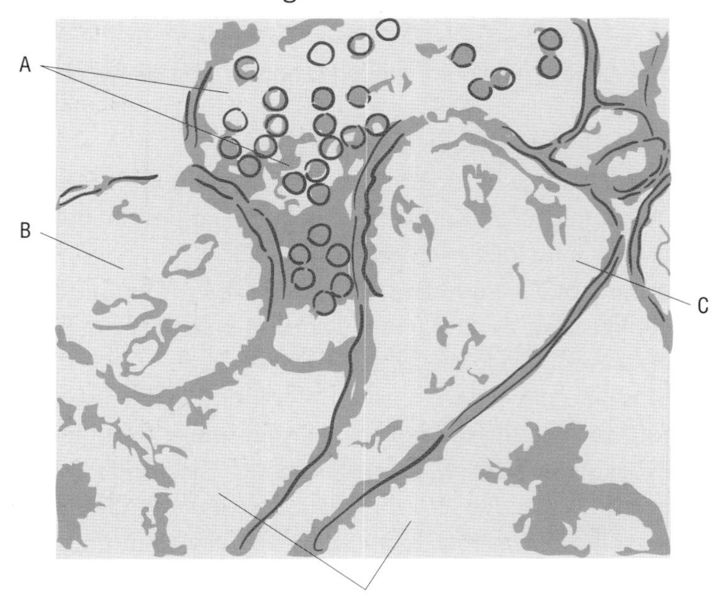

Cytoplasm of glial cells that surround and support neurones in the brain

Examiner tip

This drawing will not be quite like any of the diagrams you are familiar with. You are being asked to apply your knowledge.

(e) **(i)** Name the regions of contact. (1)

(ii) State why the structures shown would not be visible in a micrograph produced using a light microscope. (1)

(f) Use the drawing to describe how nerve impulses in A may result in nerve impulses in B and C. (6)

3 The drawing is of a small region of the glomerulus of a kidney. It shows a large cell called a podocyte, holding and supporting a loop of capillary. Cytoplasmic processes from the podocyte are shown holding the capillary, but leaving a network of spaces, so that the surface of the capillary beneath is in contact with the solution in the Bowman's capsule.

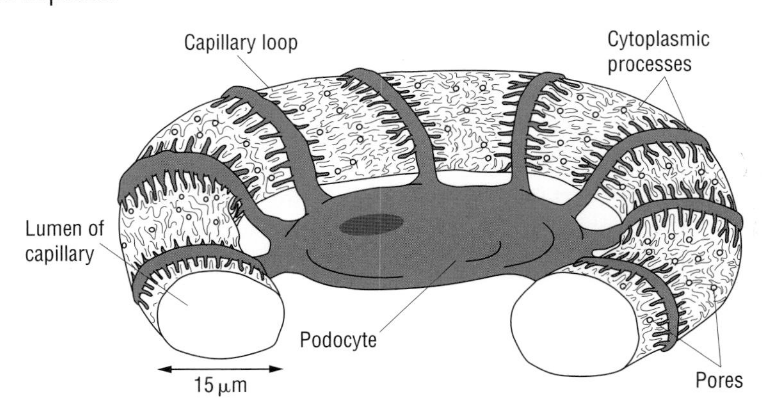

(a) Assuming the diameter of the capillary is 15 μm, calculate the approximate magnification of the drawing. (2)

(b) **(i)** Suggest why it is important that glomerular capillaries are strongly supported by the processes of podocytes. (2)

(ii) Explain why pores are needed between the fine processes of the podocytes. (1)

(c) Describe how pressure filtration is achieved by the glomerulus. (4)

4 Read the passage and answer the questions that follow.

A human kidney has about a million tubules or nephrons. Each of these tubules passes from the cortex of the kidney into the medulla. They then bend sharply round and pass back up into the cortex once more. This part of each tubule is called the loop of Henle. The walls of the descending parts of loops of Henle are permeable to ions, but not to water. Sodium and chloride ions diffuse into these descending tubules, so that the solution within becomes more and more concentrated as it passes deeper and deeper into the medulla. After the bend in the loop, this concentrated solution passes upwards towards the cortex. As it passes upwards, the walls of each tubule actively pump chloride ions out into the surrounding tissue fluid of the medulla. Sodium ions follow the chloride ions passively. The active transport of sodium chloride out of the ascending parts of the loops of Henle produces a concentration gradient of ions between the tissue fluid and the fluid in the descending tubes. It is this concentration gradient that allows sodium and chloride ions to diffuse into the descending tubes. Any individual ion may be carried around a loop of Henle many times before it 'escapes' into the distal convoluted tubule. This arrangement is called a counter-current multiplier system.

(a) Explain why the cells in the walls of the ascending parts of the tubes have many more mitochondria than the cells in the walls of the descending parts. (2)

Several nephrons drain into a collecting duct, which passes urine through the medulla and into the ureter via the renal pelvis. The permeability of the walls of the collecting ducts to water varies.

(b) Describe how the permeability of the collecting ducts is controlled and decreases after a person drinks water. (3)

(c) Suggest why a desert gerbil would have longer loops of Henle than a water vole of similar size. (2)

There is a steady net loss of sodium chloride ions from the loops of Henle to the medulla, maintaining the low water potential of the medulla.

(d) State the final destination of the ions in the tissue fluid of the medulla. (2)

Examiner tip

Don't panic when you come across a passage in the exam. Such passages are intended to help you, and it is important you study them carefully before you answer the associated questions.

5 The figure shows the metabolic pathway by which urea is formed in the mammalian body.

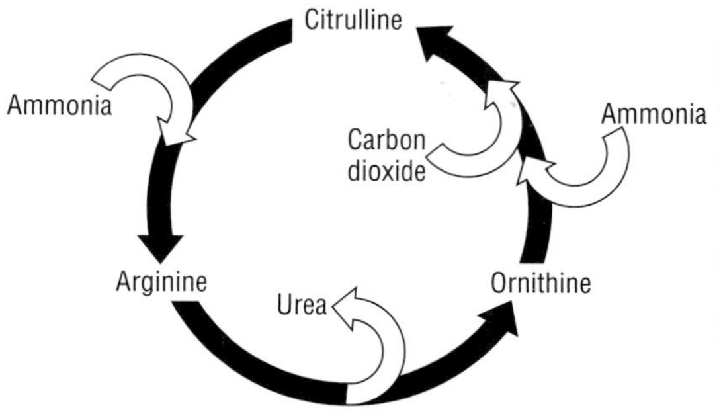

(a) (i) Name this pathway, and the organ in which it takes place. (2)

(ii) State the difference between urea and urine. (1)

(iii) Explain why a diet that included lots of meat would produce more urea than a diet mainly of fruit and potatoes. (2)

In the metabolic pathway shown in the figure, three molecules of ATP are hydrolysed. If ammonia were excreted rather than urea, this ATP would not be used.

(b) State two advantages of using urea rather than ammonia as a product for excretion. (2)

Kangaroos are herbivores that are able to survive on dead grasses, which consist almost entirely of cellulose. Kangaroos are unable to digest cellulose. The cellulose passes into the stomach, where many bacteria produce enzymes that break down and ferment the cellulose. The bacteria excrete fatty acids as a waste product of their fermentation. Much of the urea produced by a kangaroo is secreted in the saliva, rather than being excreted in the urine. The saliva is swallowed and the urea is utilised by the bacteria for protein synthesis. As the population of bacteria grows, some of them pass into the small intestine, where they are digested.

(c) Explain why a female kangaroo:

(i) can obtain energy from a diet of dead grass (1)

(ii) but would not be able to breed on this diet. (3)

Examiner tip

You will need ideas from AS unit F211 and the information provided by the passage.

6 *Pleurococcus* is a genus of green unicellular algae that grow on damp surfaces, such as stones or tree trunks. A dense suspension of *Pleurococcus* cells was prepared in a very dilute solution of potassium hydrogen carbonate ($KHCO_3$). The suspension was taken into a syringe, which was attached to a capillary tube, as shown in the figure.

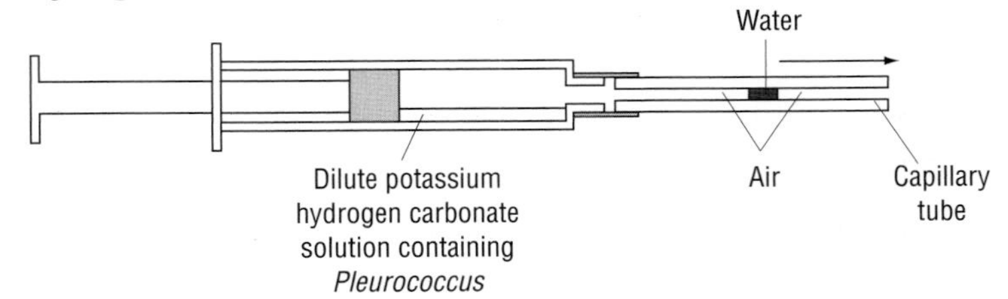

The temperature of the syringe remained constant during the experiment. When the apparatus was illuminated using a lamp, the small bead of water in the capillary tube moved steadily away from the syringe in the direction shown by the arrow.

(a) Explain why the bead of water moved in the direction indicated. (2)

(b) Describe and explain the movement of the bead of water if the apparatus was in darkness at the same temperature. (2)

(c) Describe and explain how the rate of movement of the bead would change:

(i) if the syringe was brightly illuminated and the temperature varied between 2 and 20 °C;

(ii) if the syringe was illuminated by very dim light and the temperature varied over the same range. (5)

The hydrogen carbonate ions in the solution could be made radioactive using carbon-14 (^{14}C). Under these conditions, the cells would produce many types of radioactive organic molecules in a short time.

(d) (i) Explain how it is possible for many different types of radioactive molecule to be produced. [2]

(ii) Name the first organic substance that would become radioactive in the cells. [1]

Examiner tip

This question is partly synoptic. You learnt about leaves in AS Unit 1.

7 **(a)** Explain how each of the features of a palisade mesophyll cell listed below enables the cell to maximise photosynthetic efficiency:

(i) chloroplasts which can be moved around the cell

(ii) long thin shape **(iii)** vertical orientation. (3)

Leaves of many plant species develop differently, depending on whether they grow in sunny or in shady conditions. Cuttings from the same ***Impatiens*** plant were allowed to grow at two different light intensities, in controlled environment chambers, until both cuttings had developed new leaves.

(b) State *two* variables that would have been kept constant in the controlled environment chambers. (2)

(c) Explain why cuttings from a single plant were used, rather than seedlings. (1)

Equal masses of leaves from each cutting were placed in an apparatus that allowed the volume of oxygen used or produced to be measured. Measurements were made at different light intensities and the results are shown in the graph.

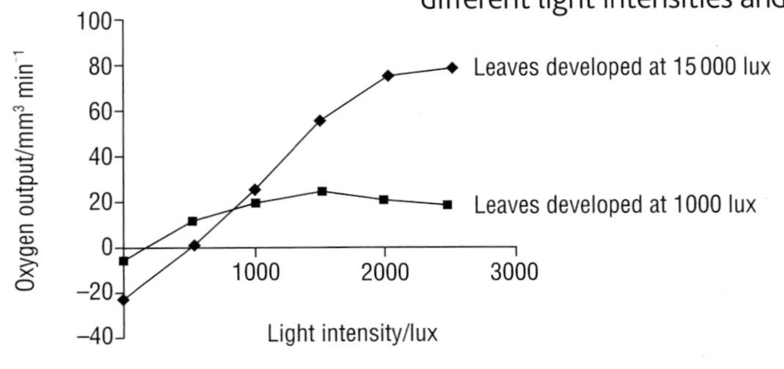

(d) With reference to the graph:

(i) explain why both sets of leaves are net absorbers of oxygen at low light intensity (1)

(ii) suggest why the leaves that developed at a light intensity of 1000 lux both produced and consumed less oxygen than the leaves that developed at 15 000 lux. (4)

8 The fixation of carbon in photosynthesis can be summarised by the equation:

$$\mathbf{3CO_2 + 9ATP + 6\ reduced\ NADP \rightarrow triose\ phosphate + 8P_i + 9ADP + 6NADP + 3H_2O}$$

(a) With reference to this equation:

(i) explain the origin of the ATP and reduced NADP (3)

(ii) explain why P_i, ADP and NADP do not accumulate in the chloroplasts during carbon fixation. (1)

(b) Explain why ribulose bisphosphate (RuBP) does not feature in the summary equation. (1)

The product of photosynthesis is often glucose or starch.

(c) Explain briefly how these molecules are produced. (2)

9 The diagram represents the Calvin cycle. The inputs to and outputs from the cycle are emphasised.

(a) With the help of the diagram, explain why a sudden fall in CO_2 concentration leads to a brief rise in RuBP concentration and a fall in GP concentration. (3)

(b) When light intensity is suddenly reduced, RuBP concentration falls and GP concentration rises. Explain why. (3)

Reduced NADP and ATP

CO_2

TP used for biosynthesis

RuBP

ATP

Examiner tip

This version of the Calvin cycle will be unfamiliar. Take plenty of time to work out what is different about it before you answer any questions.

10 The diagram shows a small part of the internal structure of a chloroplast.

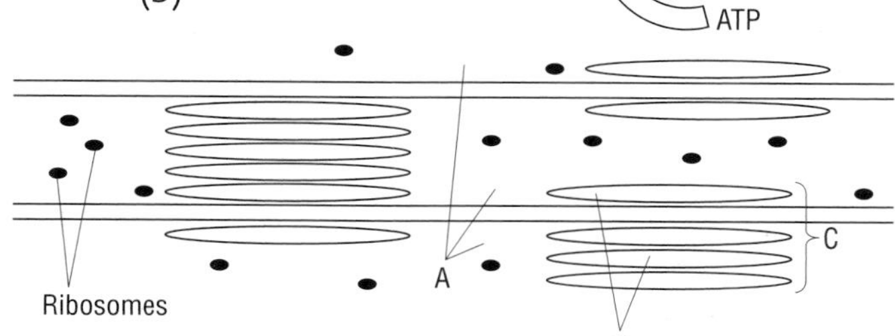

(a) (i) Name the features A, B and C. (3)

(ii) On a copy of the diagram, use the letter specified to show where each of the following would take place or be located. (6)

protein synthesis	**P**	photophosphorylation	**Z**
many photosynthetic pigment molecules located	**M**	ribulose bisphosphate carboxylase located	**R**
photolysis of water	**W**	the Calvin cycle	**Y**

Some dyes, e.g. methylene blue, are decolourised when they undergo reduction in a chemical reaction. If the molecules are re-oxidised, the colour returns. Chloroplasts that have been isolated from leaf cells and are free of the leaf cell cytoplasm are able to decolourise methylene blue but **only** in the presence of light.

(b) (i) Suggest an explanation for the ability of chloroplasts to decolourise methylene blue. (2)

(ii) Explain why blue or red light would have the greatest effect. (1)

When plant cells are viewed using a light microscope, chloroplasts can often be seen to move around in the cytoplasm of the cell.

(c) (i) Explain why this movement could not be seen using an electron microscope. (1)

(ii) Suggest an advantage for the plant of chloroplast movements within cells. (1)

11 Liver from a freshly killed mammal was quickly chilled to 2 °C. It was macerated with a blender in cold sucrose buffer solution at pH 6.8. The resulting material was passed through a fine gauze to remove cells that had not been broken open by the blender.

The resulting suspension of organelles was centrifuged at a low speed for a short time. Some material formed a precipitate that was carefully separated from the remaining suspension. The remaining suspension was centrifuged at a faster speed and for a longer time, to give a precipitate of smaller organelles with lower density. The centrifugation was repeated using faster speeds and longer times. The biological activity of the precipitates produced at each stage was investigated. The conclusions of the whole investigation are summarised in the table below. The table is incomplete.

(a) Suggest why:

(i) the liver was quickly chilled to 2 °C

(ii) a buffer solution was used

(iii) sucrose was added to the buffer. (3)

Examiner tip

You may need to revise Unit 1, Module 1 of your AS level course.

Cell fraction	Organelles in precipitate	Biological activity
A: precipitate obtained after slow speed, short time		RNA synthesis
B: precipitate obtained after moderate speed, intermediate time	Mitochondria	
C: precipitate obtained after fastest speed, longest time		Protein synthesis
D: solution remaining after centrifugation at fastest speed for longest time	No organelles	

(b) With reference to the table:

(i) name the organelles that would be expected in large numbers in precipitates A and C (2)

(ii) what is the biological activity carried out by mitochondria? (1)

(iii) explain why the solution D would be expected to carry out the reactions of glycolysis. (1)

The biological activity of cell organelles can be affected by many factors.

(c) Copy and complete the table below. The first row has been done for you. (8)

Treatment given to cell fraction B	Effect on rate of oxidative phosphorylation	Suggested explanation
Add ADP	Increase	ADP concentration may be a limiting factor for oxidative phosphorylation
Increasing temperature from 5 to 15 °C		
Add glucose		
Add pyruvate		
Bubble oxygen through the fraction		

12 Ethanol is a major constituent of wine and many other drinks.

(a) Describe how ethanol is produced in a yeast cell. (3)

Ethanol is transported by the blood from the stomach to the liver. Liver cells contain an enzyme, alcohol dehydrogenase, which oxidises ethanol, using NAD as a coenzyme.

$$\underset{\text{ethanol}}{CH_3CH_2OH} + NAD \xrightarrow[\text{(alcohol dehydrogenase)}]{} \underset{\text{ethanal}}{CH_3CHO} + \text{reduced NAD}$$

(b) (i) Suggest *two* reasons why ethanol is very rapidly absorbed from the stomach into the blood. (2)

(ii) Describe how the reduced NAD, resulting from the action of this enzyme, is used in the metabolism of the liver cell. (2)

Liver cells remove many different toxic substances from the blood apart from ethanol.

(c) Give an example. (1)

13 The diagram shows the flow of electrons through the electron transport proteins of the inner mitochondrial membrane. The electrons come from reduced NAD and reduced FAD.

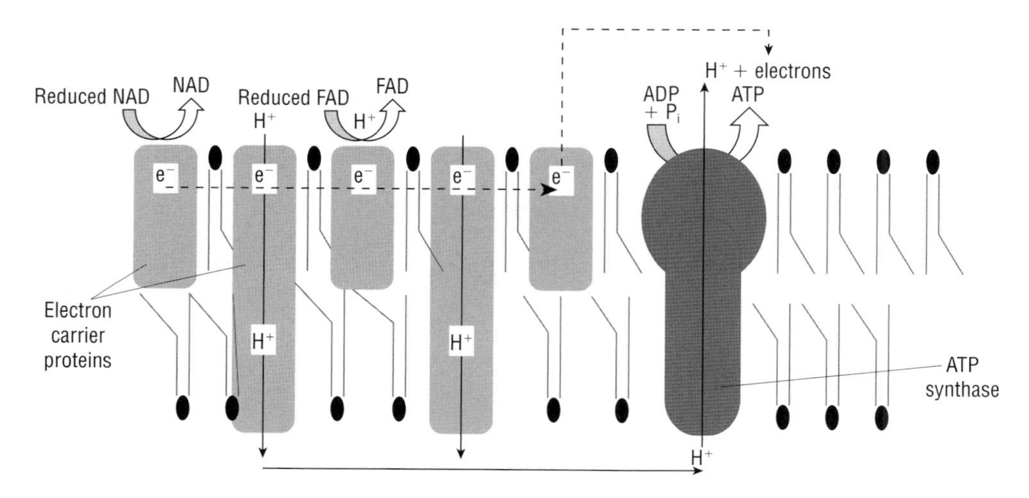

(a) Describe how the inner mitochondrial membrane is adapted to maximise the rate of ATP synthesis. (3)

The movement of hydrogen ions through the electron transport proteins is against their concentration gradient, and is therefore an example of active transport.

(b) This example of active transport is unusual. Explain how. (1)

(c) Name the process by which ATP is synthesised by the inner mitochondrial membranes. (1)

(d) After passing through ATP synthase:

(i) describe what happens to the hydrogen ions (3)

(ii) name the space/compartment/solution into which they pass. (1)

(e) Name two metabolic pathways that reduce the NAD molecules. (2)

(f) (i) Name a cell organelle, other than a mitochondrion, which contains ATP synthase. (1)

(ii) Name a process that synthesises ATP, which does not involve ATP synthase. (1)

It was originally believed that one or more of the electron carrier proteins directly synthesised ATP using energy obtained as it was alternately oxidised and reduced.

(g) Explain why this is no longer believed. (1)

Cellular control: protein synthesis

Module 1

Key words

- ribosome
- messenger RNA (mRNA)
- transcription
- translation
- codon
- transfer RNA (tRNA)
- anticodon

Examiner tip

The primary structure of a protein is its sequence of amino acids – see Unit 2 of your AS course.

✓Quick check 1

Hint

mRNA is made by assembling nucleotides when a transcript of a gene is needed.

Each group of three bases in mRNA, or 'triplet', codes for a specific amino acid. These triplets are called **codons**.

Note that U replaces T in RNA.

There is a huge amount of information stored in the DNA of our chromosomes. Each cell has a full complement of chromosomes with all the genetic information needed to make the whole body. However, individual cells need to retrieve only a small quantity of this information. For example, cells in our salivary glands use the gene that codes for the enzyme amylase. Liver cells use the gene that codes for catalase. Each gene is a code for the primary structure of a polypeptide; the sequence of bases in DNA determines the sequence of amino acids in the polypeptide, which is made by **ribosomes** in the cytoplasm. Cells have thousands of ribosomes that use short-lived transcripts of the genes in the form of **messenger RNA (mRNA)**. The process of protein synthesis occurs in four stages:

- **transcription** of DNA to make messenger RNA
- movement of mRNA from nucleus to cytoplasm
- amino acid activation
- **translation** of mRNA to make a polypeptide.

Transcription

DNA 'unzips' along the length of the gene so that the enzyme RNA polymerase can match free RNA nucleotides to form a molecule that is complementary to the template strand of the DNA molecule. This follows the rules of base pairing. The other DNA strand takes no part in the process.

Movement of mRNA to ribosomes

When transcription is finished, the messenger RNA molecule moves from the DNA to a ribosome in the cytoplasm. In a eukaryotic cell, mRNA moves from the nucleus where transcription occurs, through nuclear pores, to ribosomes.

Amino acid activation

Enzymes attach amino acids to their specific **transfer RNA (tRNA)** molecule. This needs energy supplied by ATP. The **anticodon** is a triplet of bases forming part of the tRNA molecule; it is complementary to a codon.

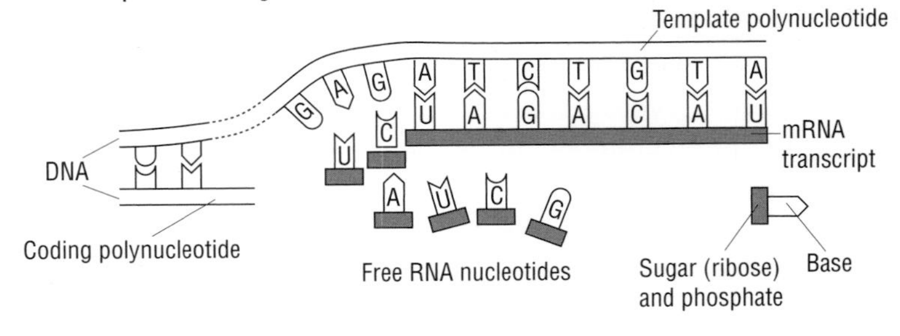

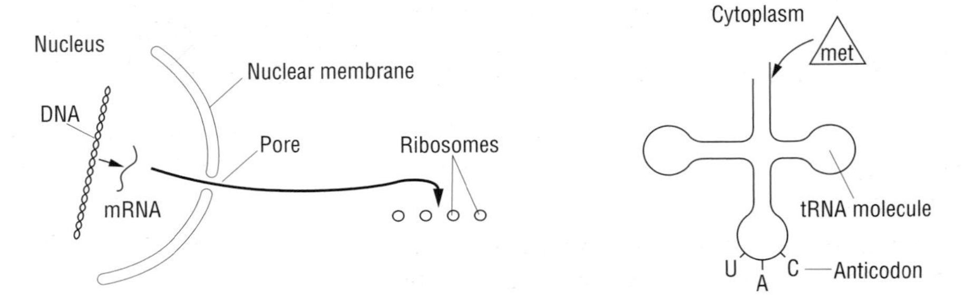

Translation

1 The mRNA molecule binds to a ribosome and translation begins. The first codon is the start codon (AUG), which codes for the amino acid methionine. The anticodon (UAC) on the tRNA molecule forms base pairs with the codon on mRNA.
2 Another tRNA molecule (here with serine) occupies the second site in the ribosome. A peptide bond forms between methionine and serine.
3 The ribosome moves one codon along the mRNA. The tRNA for methionine leaves and another tRNA (here with alanine) occupies the vacant position. Notice how tRNA molecules with anticodons that match the mRNA codons ensure that the genetic message is read correctly.
4 As the ribosome moves along the mRNA molecule, more amino acids are added to the end of the polypeptide. This carries on until the ribosome reaches a stop codon (UAA, UAG or UGA). There are no tRNA molecules for these codons, so the polypeptide breaks loose from the ribosome and translation is complete.

Examiner tip

You do not have to remember any of the triplet codes, but by following the rules of base pairing you should be able to work out the mRNA codons and tRNA anticodons from information you are given.

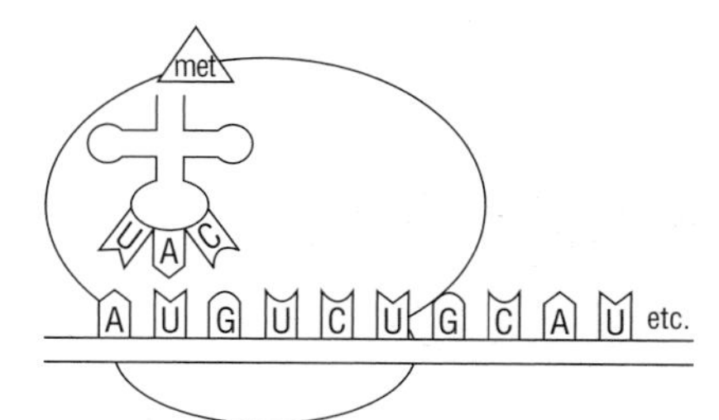

There are 20 different amino acids used to make proteins, and 64 different arrangements of the four bases into codons. This means that many amino acids have several codons and several types of tRNA.

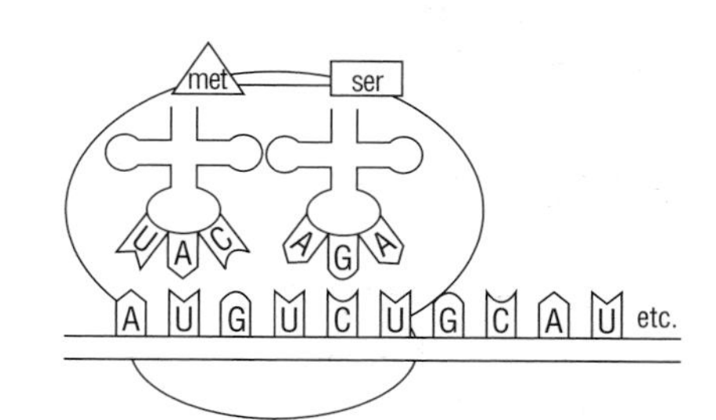

There are more codons than there are amino acids. There are four RNA codons for glycine: GGG, GGU, GGA and GGC.

DNA	RNA codon	tRNA anticodon	Amino acid
ATG	AUG	UAC	Methionine (start)
TCT	UCU	AGA	Serine
GCA	GCA	CGU	Alanine
GGG	GGG	CCC	Glycine
GTC	GUC	CAG	Valine
TAA	UAA	None	None
TAG	UAG	None	None
TGA	UGA	None	None

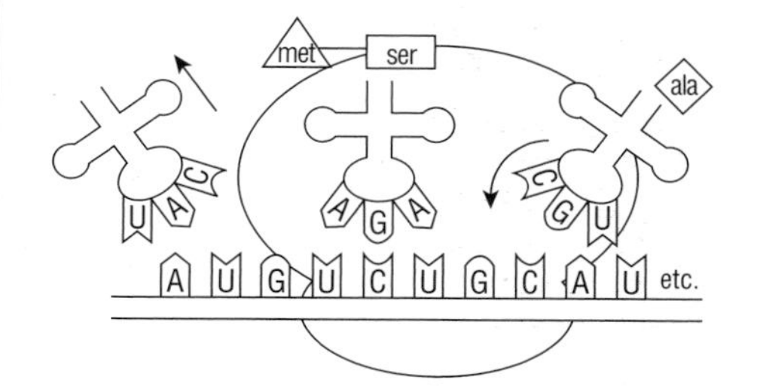

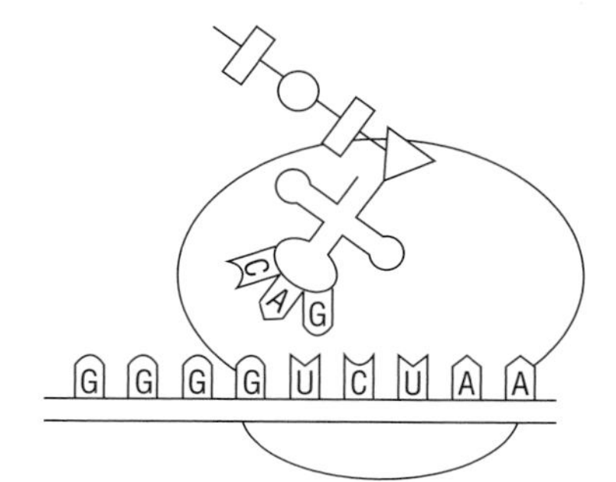

Quick check 2 and 3

Quick check 4 and 5

QUICK CHECK QUESTIONS

1 Distinguish between transcription and translation.
2 Explain why four bases can give rise to 64 different triplet codes.
3 Describe the roles of DNA, RNA polymerase, mRNA, ribosomes and tRNA in the synthesis of a polypeptide.
4 The following are codons: GGA (glycine), UCG (serine) and AAG (lysine). What are the tRNA anticodons?
5 Explain why there are 64 different codons, but only 61 different anticodons.

Mutation and the phenotype

Key words

- allele
- gene mutation
- chromosome mutation
- mutagen
- stop triplet
- sickle-cell anaemia

A mutation is an unpredictable change in the genotype of a cell that results from damage to, or imperfect replication of, the DNA, or from faulty operation of mitosis or meiosis. Such a change may be a **gene mutation** or a **chromosome mutation**.

- Mutation is a very rare event. Most genes have less than one chance in a million of being replicated incorrectly.
- Mutation takes place in a single cell, but cannot usually be detected until the mutant cell has replicated and a distinctive phenotype has been produced.
- Some chemicals, **mutagens**, increase the chance of mutation because they react with DNA or interfere with replication.
- Some types of radiation, such as X-rays, cause mutation by damaging DNA.

Examiner tip

Review the structure of proteins from Unit 2 of your AS course.

A gene mutation affects the nucleotide sequence of a section of DNA so that the triplet code for the primary structure (the amino acid sequence) of the polypeptide coded by the gene is changed.

- One or more nucleotides may be *deleted* from, or *added* to, the sequence. This alters the reading frame of the triplet code, altering all subsequent triplets.
- One nucleotide may be *substituted* for another, for example guanine instead of thymine, producing a different triplet code which may code for a different amino acid or be a **stop triplet**.
- Some substitutions in a DNA template strand do not alter the primary structure of the protein. Changing CAA to CAG, CAT or CAC has no effect. All four triplets code for valine. The mutation is 'silent'.
- A triplet sequence may be repeated many times (a 'stutter').
- Mutations may have neutral, harmful or beneficial effects on the way a protein functions.

✓*Quick check 1 and 2*

✓*Quick check 3 and 4*

A chromosome mutation alters the *number* of chromosomes in a cell or the *structure* of chromosomes, altering the number of alleles of a gene, or the sequence of genes in one or more chromosomes.

Examiner tip

Look up the structure of haemoglobin from Unit 2 of your AS course.

The genetic disease **sickle-cell anaemia** is an example of a substitution mutation with a harmful effect. All the symptoms of sickle-cell disease result from one nucleotide substitution in the triplet code for one amino acid. Close to the end of the gene coding for the β-globin chain in haemoglobin, a triplet in the normal **allele** codes for the amino acid glutamic acid. In the mutated allele, the coding is for the amino acid valine.

triplet code of template strand for Glu
CTT
normal β-globin chain .... Lys-Glu-Glu-Pro-Thr-Leu-His-Val (end of chain)
sickle-cell β-globin chain .... Lys-Glu-Val-Pro-Thr-Leu-His-Val
CAT
triplet code of template strand for Val

The presence of valine, which has a hydrophobic R group, allows the sickle cell haemoglobin (HbS) to bind to adjacent HbS molecules. Many HbS molecules can stick together in long tubules in the red cell, distorting the cell into a sickle shape. Sickle cells carry less oxygen. They get stuck in capillaries, obstructing blood flow, and are destroyed by phagocytes. Many tissues and organs of the body have a severe lack of oxygen: the person suffers from sickle-cell anaemia.

✓*Quick check 5*

The genotypes and phenotypes with respect to this gene are shown below. H^N is the allele for normal β-globin and H^S is the allele for sickle-cell β-globin.

Genotype	Phenotype
H^NH^N	Normal
H^NH^S	Sickle-cell trait: some normal and some sickle-cell haemoglobin – problems may arise during very strenuous exercise or at high altitude
H^SH^S	Sickle-cell anaemia

The presence of the allele H^S in the genotype affects susceptibility to one type of malaria: the heterozygote has a selective advantage in malarial areas compared with the homozygote H^NH^N. The malarial parasite (*Plasmodium*) does not thrive in red blood cells where some or all of the haemoglobin has sickle-cell β-globin strands. The natural selection occurring here is an example of stabilising selection (see page 58).

In contrast to the mutation resulting in sickle cell anaemia, a single substitution in the bacterial chromosome of *Escherichia coli* has a beneficial effect for the microorganism. Changing the triplet code TTT (coding for lysine) of the template strand of DNA either to TGT (threonine) or to TTG (asparagine) gives the bacterium resistance to the antibiotic streptomycin. Although the rate of growth is slightly slower than in the original, antibiotic-sensitive form, it can survive and divide in the presence of streptomycin. The antibiotic-resistant bacteria have a selective advantage in the presence of streptomycin, which acts as the selective agent. This is another example of natural selection (see page 58).

Examiner tip

Note that the symbols for the alleles follow the convention for codominant alleles.

QUICK CHECK QUESTIONS

1. State what is meant by a 'gene mutation'.
2. Explain the consequences of changing a triplet code in the middle of a gene into a stop triplet.
3. Explain how the deletion or addition of one nucleotide into the triplet code of DNA affects polypeptide production.
4. Explain why the substitution of one nucleotide for another in a length of DNA may have no effect on the polypeptide produced.
5. Describe the consequences of a person inheriting the genotype H^SH^S.

Regulation of gene activity

Key words

- β-galactosidase
- transcription
- *lac* operon
- RNA polymerase
- cyclic AMP (cAMP)
- homeotic (*Hox*) genes
- homeobox
- apoptosis

Examiner tip

Review the structure, formation and hydrolysis of disaccharides from Unit 2 of your AS course.

Hint

The action of lactose is similar to the inhibition of an enzyme by distortion of the molecule.

Quick check 1

Quick check 2

Quick check 3

The enzyme **β-galactosidase** hydrolyses the disaccharide lactose to the monosaccharides glucose and galactose.

$$\textbf{lactose + water} \xrightarrow{\beta\textbf{-galactosidase}} \textbf{glucose + galactose}$$

In the bacterium *Escherichia coli*, the number of molecules of the enzyme present in the bacterial cell varies according to the concentration of lactose in the medium in which the bacterium is growing. The bacterium has one copy of the gene for β-galactosidase, so must regulate the **transcription** of that gene in order to alter the concentration of the enzyme in its cytoplasm.

A length of DNA adjacent to the gene for β-galactosidase contains operator and promoter regions, as shown in the figure. The sequence is known as the ***lac* operon**. A regulator gene is not part of the operon and is some distance from it.

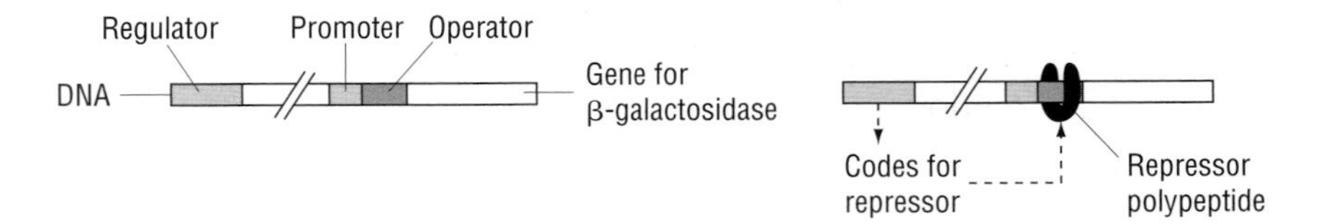

The sequence of events in the *absence* of lactose in the bacterial growth medium is as follows:

- the polypeptide coded by the regulator gene is a repressor
- the repressor binds to the operator region close to the β-galactosidase gene
- in the presence of bound repressor at the operator, **RNA polymerase** cannot bind to the DNA at the promoter region
- no transcription of the gene for β-galactosidase occurs.

When lactose is *present* in the bacterial growth medium:

- lactose is taken up by the bacterium
- lactose binds to the repressor molecule, distorting its shape and preventing it from binding to DNA at the operator site
- transcription is no longer inhibited and β-galactosidase is produced – the gene is switched on.

In this way, the bacterium produces β-galactosidase only when lactose is present in the surrounding medium. It avoids the waste of energy in producing an enzyme for hydrolysing a sugar it may never encounter. But when the sugar is available, it can be hydrolysed.

When glucose is available, it is used in preference to other sugars. So when *E. coli* finds both glucose and lactose in its surroundings, it represses the use of lactose by suppressing the *lac* operon. This occurs through the use of a transcription control protein that is active *only* in the presence of **cyclic AMP (cAMP)**. Remember, cAMP alters the activity of proteins by altering their three-dimensional structure (see page 47). Glucose *reduces* the amount of cAMP in the bacterial cell.

Cyclic AMP

Some proteins that are needed for transcription to occur are activated by cAMP, which alters their three-dimensional structure. An example is catabolite activator protein (CAP) in bacteria. The protein has a region that binds to DNA and a region that activates transcription. In the *absence* of cAMP, CAP cannot bind to DNA, in turn preventing RNA polymerase from starting transcription.

✓ Quick check 4

The homeobox

The genes that control the development of body plans are similar in plants, animals and fungi. The so-called **homeotic** (***Hox***, or homeobox encoding) **genes** are expressed in specific patterns and at particular stages of development, when activated by proteins such as sonic hedgehog protein. The name 'sonic hedgehog' is taken from an animated cartoon character. The 'hedgehog' proteins are vital for pattern formation.

The **homeobox** is a sequence of DNA that codes for a region of 60 amino acids found in proteins of many, or even all, eukaryotes. This region is able to bind to DNA so that the proteins can regulate transcription. In animals, the homeobox is common in genes concerned with the control of developmental events such as segmentation, the establishment of an anterior–posterior axis, and the activation of genes coding for body parts such as limbs.

Examiner tip

Remind yourself of what is meant by a eukaryote from Unit 2 of your AS course.

✓ Quick check 5

Apoptosis

Apoptosis, or programmed cell death, can be triggered to change a body plan. As a developing mammalian limb extends, *Hox* genes play an important part in organising the skeleton. However, the final structure of the limb depends on the regulation of cell death, for example to separate the ulna and radius bones in the forearm (see page 85) and to form joints. Cell death also occurs in soft tissue. The fingers of a developing human hand are joined by webs of tissue. The fingers are then separated by apoptosis of the joining tissue. In apoptosis, the cell shrinks and chromatin condenses at the edge of the nucleus. The DNA of condensed chromatin is not transcribed.

✓ Quick check 6

QUICK CHECK QUESTIONS

1. Explain why preventing RNA polymerase from binding to DNA prevents transcription.
2. State the advantages of a mechanism (such as the *lac* operon) for switching on a gene only in certain circumstances.
3. Explain how the presence of glucose in a bacterial cell suppresses the *lac* operon.
4. Describe how a lack of cAMP can prevent transcription.
5. State what is meant by the term 'homeobox'.
6. Describe one role of apoptosis in changing a body plan.

Meiosis

Module 1

Key words

- haploid (n)
- prophase
- homologue
- synapsis
- bivalent
- crossing-over
- metaphase
- independent assortment
- anaphase
- telophase
- genetic variation

Examiner tip

You should remind yourself of mitosis, meiosis and homologous chromosomes from Unit 1 of your AS course before tackling further details of meiosis.

Hint

A bivalent consists of *two* chromosomes and therefore *four* chromatids.

Hint

A microtubule is a hollow cylinder made of the protein tubulin.

Examiner tip

Review the cytoskeleton from Unit 1 of your AS course.

Meiosis is the process by which a nucleus divides by two divisions into four nuclei, each containing half the number of chromosomes of the mother cell. The resulting nuclei are **haploid (n)**. Meiosis is necessary for sexual reproduction in eukaryotes; without it fertilisation would double the chromosome number every generation.

Division I

In **prophase I**, homologous chromosomes (**homologues**), each consisting of two genetically identical sister chromatids held together by their centromere, form pairs in the process of **synapsis**. Each pair is called a **bivalent**. **Crossing-over** occurs between the homologous chromosomes of a bivalent (see opposite page), creating genetic variation. The chromosomes condense and become more visible, as in mitosis. At the end of prophase I, the nuclear envelope breaks down. Centrioles, if present, duplicate and move to opposite poles. A spindle of microtubules forms.

In **metaphase I**, bivalents line up independently on the cell's equator and each homologue attaches to spindle microtubules by its undivided centromere. The orientation of the maternal and paternal chromosomes of each bivalent is random. This **independent assortment** of paternal and maternal chromosomes gives further genetic diversity.

Homologues, each consisting of two chromatids that are no longer genetically identical because of crossing-over, are moved apart to opposite poles by the shortening of the microtubules of the spindle in **anaphase I**.

The chromosomes reach opposite poles of the cell in **telophase I**. After this, new nuclear envelopes form and the cell may divide into two haploid cells.

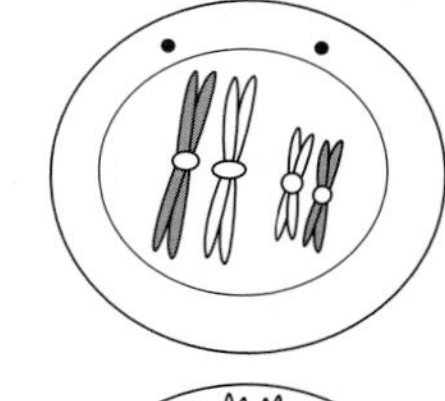

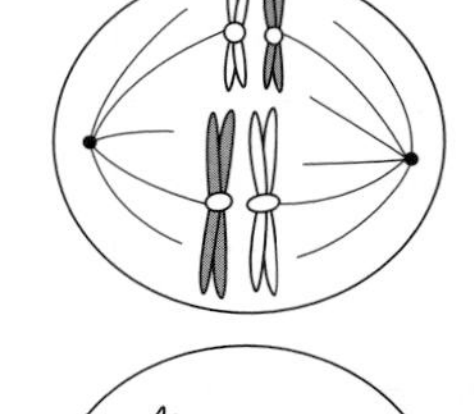

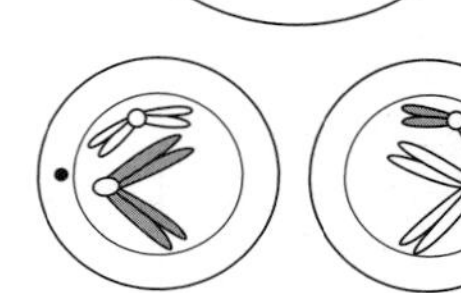

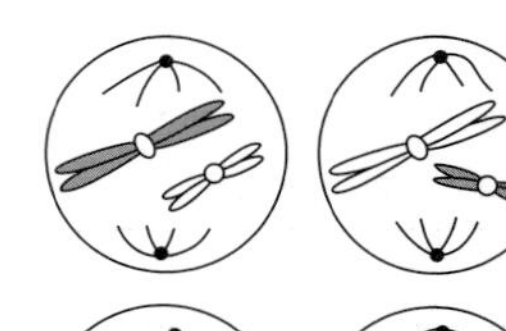

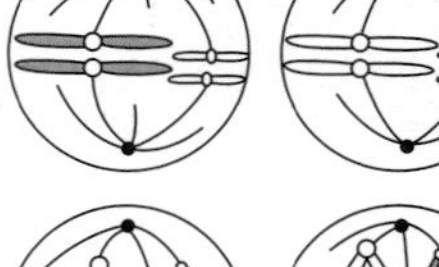

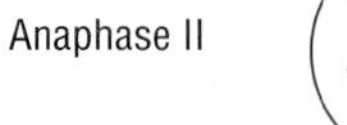
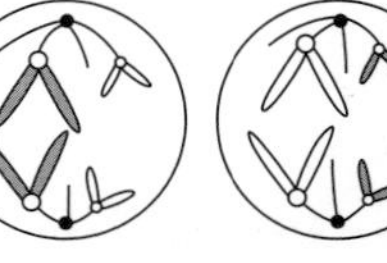

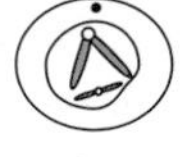

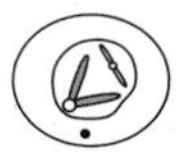
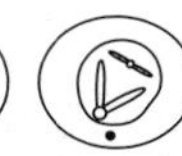

Division II

In prophase II, centrioles, if present, duplicate and move to the poles. A new spindle forms. The nuclear envelopes break down.

The chromosomes are moved to the equator in metaphase II and attach to the spindle by their centromeres, which then divide.

Then, in anaphase II, sister chromatids are separated.

Finally, in telophase II, new nuclear envelopes form and the cells divide.

In meiosis II, the chromatids of a homologue separate as in a mitotic division. The final result is four nuclei, each with half the chromosome number of the original nucleus and all genetically different from one another.

✔Quick check 1 and 2

Genetic variation

Meiosis and fertilisation lead to **genetic variation**. Genetic variation is generated in three main ways.

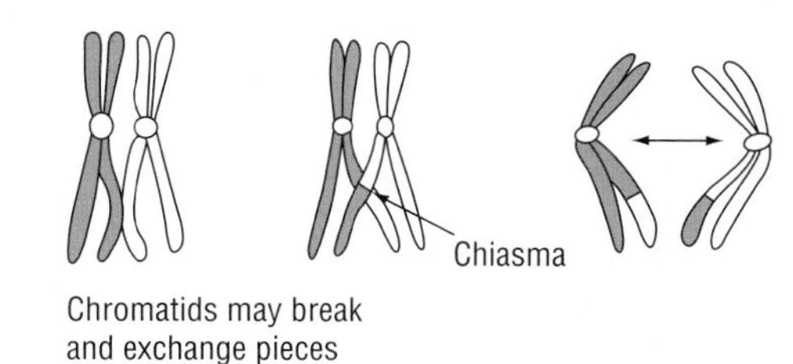

Chromatids may break and exchange pieces

- Crossing-over – during prophase I of meiosis, non-sister chromatids swap equivalent portions of chromatids, giving new combinations of alleles. In a few organisms, e.g. male *Drosophila*, no crossing-over takes place, but in most eukaryotes the number of cross-overs is proportional to the length of the chromosome.
- Independent assortment – at metaphase I of meiosis, there is a 50:50 chance which way round a pair of homologous chromosomes (a bivalent) will be placed on the equator of the cell. This gives 2^n possible combinations of paternally derived and maternally derived chromosomes in the daughter cells (where n is the haploid number of chromosomes).

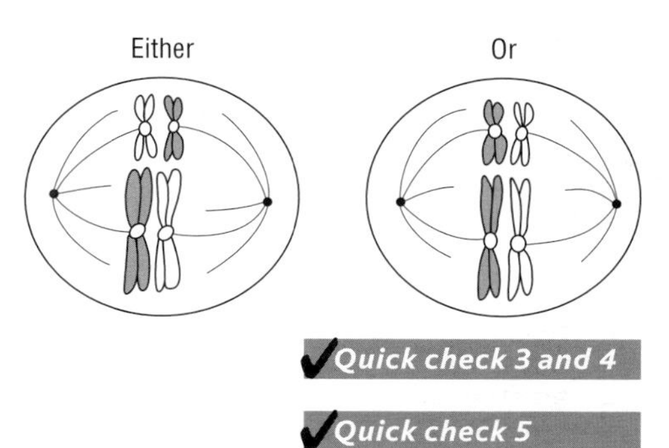

- Meiosis halves the chromosome number, which is restored during fertilisation by the *random* fusion of two gametes to give a zygote.

✔Quick check 3 and 4

It is useful to compare mitosis and meiosis.

✔Quick check 5

Mitosis	Meiosis
Maintains the chromosome number	Halves the chromosome number
Has one division cycle	Has two division cycles
Gives two daughter nuclei	Gives four daughter nuclei
Does not involve crossing-over and independent assortment	Involves crossing-over and independent assortment
Gives daughter nuclei that are genetically identical (apart from mutation) to one another and to the parent nucleus	Gives daughter nuclei that are genetically different from one another and from the parent nucleus

QUICK CHECK QUESTIONS

1. Describe the essential differences between meiosis I and meiosis II.
2. State the similarities between meiosis II and mitosis.
3. Explain what occurs in a bivalent during crossing-over.
4. In a male *Drosophila*, only independent assortment produces genetic variation in meiosis. The diploid ($2n$) chromosome number is 8. Calculate the number of genetically different spermatozoa that can be produced.
5. Draw out and complete the table below to show the different outcomes of mitotic and meiotic divisions of a cell with a diploid number of 16.

	Mitosis	Meiosis
Number of division cycles		
Number of daughter cells		
Number of chromosomes per nucleus in daughter cells		

Examiner tip

Questions 2 and 5 are synoptic. Look up mitosis from Unit 1 of your AS course.

Genetics: terminology and codominance

Module 1

Key words

- phenotype
- genotype
- dominant
- recessive
- codominant
- locus
- Punnett square

Genetics is the study of genes, their effects and their inheritance. It also encompasses DNA technology, which includes genetic manipulation in, for example, genetic engineering and gene therapy. Genomics is the attempt to map and describe all the genes of various organisms.

Genetic terminology

Gene	A length of DNA that codes for the production (via mRNA) of one or more polypeptides or codes for the direct production of rRNA or tRNA.
Allele (allelomorph)	A variant form of a particular gene. There may be two, several or very many alleles of a gene.
Genotype	The alleles present in an individual or cell.
Genome	The sum total of genes in an organism.
Phenotype	The expression of the alleles of the genotype, giving the individual's observable traits.
Homozygote	A genotype in which the two alleles of a gene are identical, e.g. AA or aa.
Heterozygote	A genotype in which the two alleles of a gene are different, e.g. Aa.
Dominant allele	A dominant allele has the same effect on the phenotype when it is heterozygous as when it is homozygous. It is symbolised by an upper-case letter, e.g. T for tall.
Recessive allele	A recessive allele affects the phenotype only when it is homozygous. It is symbolised by a lower-case letter, e.g. t for short (not tall), and must be the same letter as used for the dominant allele.
Codominant alleles	Codominant alleles both affect the phenotype in a heterozygous organism. They are symbolised by upper-case superscripts to an upper-case letter representing the gene, e.g. I^A and I^B for human blood groups A and B.
Multiple alleles	Most genes have more than two alleles; some very many.
Locus	The position of a gene on its chromosome.
Homologous chromosomes (homologues)	Homologues have the same genes in the same sequence (except for chromosome mutation), but the alleles of the genes may be different on the paternal and maternal chromosomes. Homologous chromosomes are able to form bivalents during prophase I of meiosis.
Sex linkage	Genes that are inherited together on a sex chromosome show sex-linked inheritance, e.g. the genes on the X chromosome.
F_1: first filial generation	F_1 should only be used to denote the first generation resulting from crossing two *homozygotes*. In other circumstances 'offspring 1' should be used.
F_2: second filial generation	F_2 should only be used when crossing two members of the F_1 generation. In other circumstances 'offspring 2' should be used.
Test cross	A cross to test whether an individual with a dominant phenotype is homozygous or heterozygous. The individual is crossed with a homozygous recessive individual.

Quick check 1, 2 and 3

Genetic diagrams

A genetic diagram is the accepted way of showing the genotypes and phenotypes of the parents and expected offspring of those parents.

Genotype	Phenotype
C^RC^R	Red flowers
C^RC^W	Pink flowers
C^WC^W	White flowers

The inheritance of red, pink or white flower colour in snapdragons (*Antirrhinum*) is controlled by two alleles that show codominance. Allele C^R gives red flowers and allele C^W white flowers. Three genotypes and three phenotypes are possible.

Consider a cross between a red-flowered and a white-flowered snapdragon:

Parental phenotypes	Red flowers	White flowers
Parental genotypes	C^RC^R	C^WC^W
Genotypes of gametes	All (C^R)	All (C^W)
Genotype of offspring (F_1)	All C^RC^W	
Phenotype of offspring	pink flowers	

Now consider a cross between a pink-flowered and a white-flowered snapdragon:

Parental phenotypes	Pink flowers	White flowers
Parental genotypes	C^RC^W	C^WC^W
Genotypes of gametes	(C^R) or (C^W) in equal proportions	all (C^W)

Genotypes and phenotypes of offspring:

		Gametes from white parent
		(C^W)
Gametes from pink parent	(C^R)	**C^RC^W** pink flowers
	(C^W)	**C^WC^W** white flowers

Approximately equal numbers (a ratio of 1:1) of pink- and white-flowered offspring are expected from this cross.

A *monohybrid* cross is concerned with the inheritance of *one* gene, such as the inheritance of flower colour in snapdragons shown here. A *dihybrid* cross looks at the inheritance of *two* genes.

Hint

The symbols for the alleles follow the convention for codominant alleles. Gametes are shown with circles round them.

Examiner tip

Use a **Punnett square** like this to predict the results of a cross, even in simple examples where it may seem obvious.

Examiner tip

All the gametes from the white parent are the same (C^W). Therefore use C^W only once in the Punnett square.

✓ Quick check 4

Module 1

QUICK CHECK QUESTIONS

1 What is an 'allele'?

2 Explain the terms 'homozygote' and 'heterozygote'.

3 Distinguish between dominant, recessive and codominant alleles.

4 A red-flowered and a pink-flowered snapdragon are crossed. Draw a genetic diagram to show the genotypes and phenotypes of the parents and offspring.

Genetics: sex linkage and epistasis

Key words

- sex chromosomes X and Y
- autosome
- epistatic gene
- hypostatic gene

Inheritance of sex and sex linkage

In most animal species, sex is determined by the inheritance of the **sex chromosomes: X** and **Y**. Unlike the other pairs of chromosomes (**autosomes**), X and Y chromosomes are not the same size. They are homologous for only part of their length, and this homologous region allows them to pair during meiosis. In mammals, the Y chromosome is much shorter than the X. The X chromosome carries a large number of genes, while the Y has very few. A person with a Y chromosome is male. A female (XX) has two alleles of all the genes carried on the X chromosome. A recessive allele will affect the phenotype only if the individual is homozygous for that allele. Much of the X chromosome in a male (XY) is unmatched, because the Y chromosome is so small. A recessive allele of a gene on the X chromosome can affect the phenotype because it is the *only* allele of that gene present.

One of the genes on the X chromosome, H/h, codes for the production of a blood clotting agent, factor VIII. The dominant allele, H, codes for the production of normal factor VIII, while the recessive allele, h, results in lack of the factor leading to the disease haemophilia A.

Quick check 1

Hint

The alleles of sex-linked genes are shown as superscript to an X (the chromosome) using upper case (dominant) or lower case (recessive) letters, e.g. X^H or X^h.

A haemophiliac man (X^hY) and a normal woman (X^HX^H) have children:

Parental phenotypes	Haemophiliac man	Normal woman
Parental genotypes	X^hY	X^HX^H
Genotypes of gametes	(X^h) and (Y) in equal proportions	all (X^H)

Genotypes and phenotypes of offspring:

		Gametes from woman
		(X^H)
Gametes from man	(X^h)	X^HX^h carrier female
	(Y)	X^HY normal male

Now suppose that a woman carrier of haemophilia and a normal man have children:

Parental phenotypes	Normal man	Carrier woman
Parental genotypes	X^HY	X^HX^h
Genotypes of gametes	(X^H) and (Y) in equal proportions	(X^H) and (X^h) in equal proportions

Genotypes and phenotypes of offspring:

		Gametes from woman	
		(X^H)	(X^h)
Gametes from man	(X^H)	X^HX^H normal female	X^HX^h carrier female
	(Y)	X^HY normal male	X^hY haemophiliac male

Sons inheriting X^h are haemophiliacs; daughters inheriting X^h are carriers.

✓Quick check 2 and 3

Epistasis

Epistasis arises from the interaction of different gene loci in the expression of a phenotypic character. The **epistatic gene** at one locus alters or inhibits the expression of a second locus, the **hypostatic gene**. The protein product of the epistatic gene influences or controls the expression of the hypostatic gene.

Dominant epistasis is seen in the inheritance of feather colour in poultry. Individuals carrying the dominant allele of one gene (A/a) have white feathers even when they carry the dominant allele of a second gene (B/b) for coloured feathers. Genotypes A–B–, A–bb and aabb are white. Only aaB– has coloured feathers. In a cross between two individuals each with the genotype **AaBb**, the usual expected Mendelian ratio of 9:3:3:1 is modified by epistasis to (9 + 3 + 1):3, giving 13 white:3 coloured birds.

Hint

You do not need to be able to produce genetic diagrams involving epistasis.

Recessive epistasis is seen in the inheritance of flower colour in *Salvia*. The alleles for purple (B) and pink (b) can be expressed only in the presence of a dominant allele (A) of a second gene. Genotype A–B– has purple flowers and A–bb has pink flowers. Genotypes aaB– and aabb have white flowers. In a cross between two *Salvia* plants each with the genotype AaBb, the usual Mendelian ratio of 9:3:3:1 is modified by epistasis to 9:3:(3 + 1), giving 9 purple:3 pink:4 white-flowered plants.

✓Quick check 4, 5 and 6

QUICK CHECK QUESTIONS

1 Explain why a single X-linked recessive allele can affect the phenotype of a male mammal.

2 State the genotype of a haemophiliac woman.

3 Very rarely, a female carrier of haemophilia and a haemophiliac man have children. Draw a genetic diagram to show the genotypes and phenotypes of the offspring that could be produced.

4 Explain what is meant by 'epistasis'.

5 Using the alleles given above, what are the possible genotypes of coloured poultry?

6 Using the alleles given above, what are the possible genotypes of white *Salvia*?

Genetics: χ^2 (chi-squared) test

Module 1

Key words

- χ^2 test
- probability (p)
- degrees of freedom (η)
- significance

χ^2 (chi-squared) test

The χ^2 **test** is a way of estimating the **probability (p)** that differences between observed and expected results are due to chance. The formula for χ^2 is:

$$\chi^2 = \text{the sum of } \frac{(\text{observed number} - \text{expected number})^2}{\text{expected number}} = \sum \frac{(O-E)^2}{E}$$

- Differences are squared, making all numbers positive. If the differences were simply added together, some differences would be larger than expected, giving a positive number and others would be smaller, giving a negative number. These would cancel one another out if they were added in that form.
- Dividing each (difference)^2 by the expected value is a way of allowing for the number of results.

How χ^2 is calculated

The results of a cross, in which a 1:1 ratio of phenotypes was expected, gave 43 offspring with one phenotype and 57 with the other phenotype. Is such divergence from expectation due to chance?

The calculation of χ^2 is shown below:

	First phenotype	Second phenotype
O	43	57
E	50	50
$O-E$	–7	7
$(O-E)^2$	49	49
$(O-E)^2/E$	0.98	0.98
$\Sigma(O-E)^2/E=\chi^2$	1.96	

✓Quick check 1

Now look at the table below, which shows part of a table of χ^2 values and probabilities for different **degrees of freedom**. The number of degrees of freedom, η, takes account of the number of comparisons made. It is calculated as (number of classes of data – 1). Here, the calculation involves two classes of data, so $\eta = (2–1) = 1$.

Degrees of freedom (η)	Probability (p)				
	0.1	0.05	0.02	0.01	0.001
1	2.71	3.84	5.41	6.64	10.83
2	4.61	5.99	7.82	9.21	13.82
3	6.25	7.82	9.84	11.35	16.27
4	7.78	9.49	11.67	13.28	18.47

In the $\eta = 1$ row of the table, the critical value of χ^2 is 3.84. Any value of χ^2 less than 3.84 (and here it is only 1.96) means that there is more than a 5% probability ($p = 0.05$, or 1 in 20) of a difference of 43:57 coming about by chance alone. Therefore there is no reason to suppose that the original prediction of a 1:1 ratio was wrong.

In this example, the original prediction that a 1:1 ratio of phenotypes was expected from the cross is a ***null hypothesis***. This simply means that it is a hypothesis constructed in order to give a set of expected results which may then be compared with the observed results obtained from an actual experiment. In the example above the null hypothesis can be accepted.

The larger the difference between the observed and expected results, the greater the likelihood that the original prediction was wrong. When the difference between the observed and expected results is so large that it would be expected to occur by chance in *fewer* than 1 in 20 experiments (a probability of 0.05), the result is said to differ **significantly** from expectation. The prediction on which the expected results were based is not upheld.

Examiner tip

The predicted results of a dihybrid genetic cross may assume independent assortment. Linkage or epistasis may explain results that differ from expectation.

Quick check 2, 3 and 4

QUICK CHECK QUESTIONS

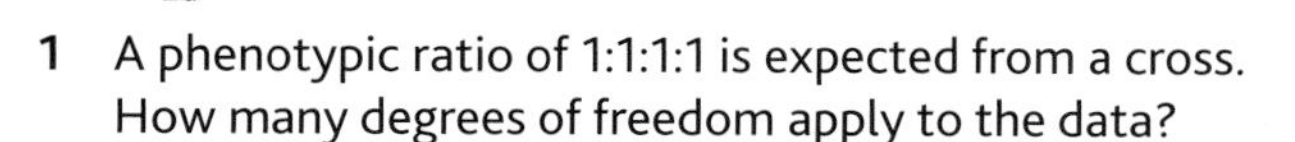

1 A phenotypic ratio of 1:1:1:1 is expected from a cross. How many degrees of freedom apply to the data?

2 In a test, the probability that the differences between the observed and expected results were due to chance was found to be 0.001. Are the differences due to chance or to some other factor?

3 The results of a cross in which a 3:1 ratio of phenotypes was expected gave 160 offspring with one phenotype and 40 with the other. Use the χ^2 test to find whether the divergence from expectation is due to chance.

4 The results of a cross, in which a 1:1:1:1 ratio of phenotypes was expected, gave the following numbers of four phenotypes: 70, 65, 35, 30. Use the χ^2 test to find out whether the divergence from expectation is due to chance, or whether the expected ratio was based on an incorrect assumption.

Examiner tip

You will always be given the formula for χ^2, or even the value. You need to know what to do with it, and what it tells you about the results.

Phenotypic variation

Key words

- phenotype
- qualitative variation
- quantitative variation
- discontinuous variation
- continuous variation
- polygenes
- phenotypic variation (V_P)
- environmental variation (V_E)
- genetic variation (V_G)
- Hardy–Weinberg principle

✓Quick check 1 and 2

The traits shown by an organism are its **phenotype**. Phenotypic differences between the readers of this revision guide will include **qualitative** differences, such as different blood groups, and **quantitative** differences, such as height or mass.

- Qualitative differences fall into clearly distinguishable categories, with no intermediates. This is **discontinuous variation**.
- With quantitative traits, there are no distinguishable categories. Instead, there is a range of values between two extremes. This is **continuous variation**.

In discontinuous (qualitative) variation:

- different alleles at a single gene locus have large effects
- different gene loci have different effects on the trait.

In continuous (quantitative) variation:

- different alleles at a single gene locus have small effects
- different gene loci have the same, often additive effect on the trait
- a large number of gene loci (**polygenes**) may have a combined effect on the trait.

Genotype and environment contribute to **phenotypic variation**. Their interaction can be written as:

$$V_P = V_G + V_E$$

where V_P = phenotypic variation; V_G = the genetic component of variation and V_E = the environmental component of variation.

Examiner tip

Revise variation, adaptations and natural selection from Unit 2 of your AS course.

Environmental effects may allow the full genetic potential to be reached, or may restrict it in some way. A pea plant with the dominant allele for tallness may grow only to the same height as a plant homozygous for the recessive allele for shortness if it is growing in soil with few nutrients.

Environmental variation (V_E) cannot be inherited: it is the result of differences in the environment experienced during an individual's lifetime. It is important to recognise the extent of environmental variation when selectively breeding organisms. Because environmental variation cannot be inherited, it cannot be selected for (see page 58).

✓Quick check 3 and 4

The allele producing curled wings (which are not suitable for flight) in *Drosophila* is expressed only if the pupae are kept at a higher temperature than normal (27 °C) for the last day of their pupal life.

In the 'Himalayan' colouring of rabbits, and in Siamese and Burmese cats, the allele for the formation of dark pigment is only expressed at low temperatures. The extremities (ears, nose and paws) are cooler than the core temperature of the body, so the allele is able to have its effect in these cooler regions.

✓Quick check 5 and 6

Variation within a population means that some individuals may have features that give them an advantage compared with other individuals with different phenotypes. This is the basis of natural selection (see page 58).

Hardy–Weinberg principle

When a particular trait is controlled by two alleles of a single gene, A and a, the population will be made up of three genotypes: AA, Aa and aa. Calculations based on the **Hardy–Weinberg principle** show the basic rules determining the proportions of each of these genotypes in a large, randomly mating population.

The *frequency* of a genotype is its proportion of the total population. The total is the whole population (i.e. 1) and frequencies are decimals (e.g. 0.25) of the total.

Use the letter p to represent the frequency of the dominant allele, A, in the population and the letter q to represent the frequency of the recessive allele, a. Then, since there are only two alleles of this gene:

$p + q = 1$ (the whole population) Equation 1

- The chance of an offspring having a dominant allele from both parents $= p \times p = p^2$
- The chance of an offspring having a recessive allele from both parents $= q \times q = q^2$
- The chance of an offspring having a dominant allele from the father and a recessive allele from the mother $= p \times q = pq$
- The chance of an offspring having a dominant allele from the mother and a recessive allele from the father $= p \times q = pq$

So, $p^2 + 2pq + q^2 = 1$ (the whole population) Equation 2

The homozygous recessives (q^2) in the population can be recognised and counted. Suppose the incidence of the aa genotype is 1 in 100 individuals (1%).

Then, $q^2 = 0.01$ and $q = \sqrt{0.01} = 0.1$

So, using Equation 1, $p = 1 - 0.1 = 0.9$ and $p^2 = (0.9)^2 = 0.81$. That is, 81% of the population are homozygous AA.

And, $2pq = 2 \times 0.9 \times 0.1 = 0.18$ (or, using Equation 2, $2pq = 1 - (0.01 + 0.81) = 0.18$). That is, 18% of the population are heterozygous Aa.

Examiner tip

The Hardy–Weinberg calculations do *not* apply when the population is small, or when there is:

- significant selective pressure against one of the genotypes
- migration of individuals carrying one of the two alleles into, or out of, the population
- non-random mating.

Quick check 7

Quick check 8

QUICK CHECK QUESTIONS

1. Distinguish between continuous and discontinuous variation.
2. Explain the genetic basis of continuous variation.
3. State what is meant by the equation $V_P = V_G + V_E$.
4. Why can't V_E be inherited?
5. Explain why the allele producing 'Himalayan' colouring in rabbits affects only the extremities of the animal.
6. Predict the phenotypes of two groups of genetically identical dwarf pea plants when one group is grown in high light intensity, and the other group is grown in low light intensity (but in otherwise identical environmental conditions).
7. A trait is controlled by a single gene with two alleles, A and a. Explain why only the homozygote recessive, aa, can be recognised.
8. Calculate the proportion of homozygous dominant individuals and of heterozygotes in a population in which the proportion of homozygous recessives is 16%.

Genetic drift and natural selection

Key words

- genetic drift
- natural selection
- founder effect
- fitness
- species
- biospecies
- phylogeny
- clade
- speciation
- allopatric
- sympatric
- polyploidy

The processes that act on a natural population of organisms to alter allele frequencies fall into two categories:

- random processes, e.g. genetic drift
- non-random processes, e.g. selection.

Genetic drift

Genetic drift is a change in allele frequency that occurs *by chance*, because only some of the organisms in each generation reproduce. Genetic drift is particularly noticeable when a *small* number of individuals are separated from the rest of a large population. They form a small sample of the original population, and so are unlikely to be representative of the large population's gene pool. Genetic drift in the small population will alter the allele frequencies still further. This process, occurring in a recently isolated small population, is called the **founder effect**.

Quick check 1

Hint

A population's gene pool is the sum total of its alleles.

Selection

Genetic variation (and so phenotypic variation) in a population of organisms means that some organisms will have a better chance of survival than others. They are better adapted to the particular conditions, and are more likely than their less fit competitors to breed successfully and contribute more offspring to the next generation. This **natural selection**:

- increases the chance of advantageous adaptive alleles being passed to the next generation
- decreases the chance of non-adaptive alleles being passed to the next generation.

Examiner tip

Review what you have learned about natural selection in Unit 2 of your AS course.

An individual's **fitness** is a measure of its ability to transmit its alleles to the next generation as a result of natural selection.

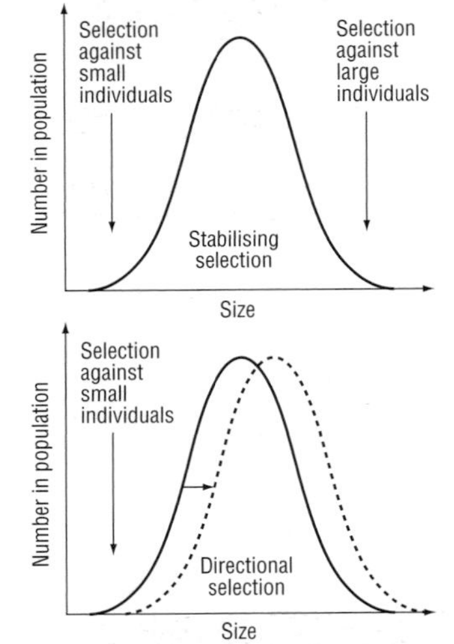

In unchanging conditions, *stabilising* selection maintains existing adaptations and so maintains existing allele frequencies. Stabilising selection is seen best where the heterozygote for a particular gene has a selective advantage and the two homozygotes have selective disadvantages. This maintains all genotypes in the population because it tends to be the heterozygotes that breed. An example is human heterozygotes for sickle cell anaemia in a malarial area.

In changing conditions, *directional* selection alters allele frequencies. An example is pollution favouring industrial melanism in the peppered moth, *Biston betularia*.

A new allele (a mutation) appearing in the population may be disadvantageous in existing conditions and may be removed by stabilising selection. But should conditions change, the new allele might be advantageous and selected for. Selection then becomes an evolutionary force.

Quick check 2

Examiner tip

Do not confuse evolutionary fitness with physiological fitness (as used in sports physiology).

Species and speciation

It is difficult to produce a definition of the term **species** that satisfies all biologists. Two definitions are commonly used.

- A group of more-or-less similar organisms capable of interbreeding and producing fertile offspring and that are reproductively isolated from other such groups. This is a **biospecies**.

- A group of organisms showing a close similarity in a number of characteristics (morphological, physiological, embryonic, ecological and behavioural) that is the basis of a classification showing its evolutionary relationships, or **phylogeny**. A phylogenetic lineage is called a **clade**.

The first definition, that of a 'good' species or biospecies, excludes:

- organisms that reproduce asexually or by parthenogenesis (the development of an *unfertilised* gamete into a new individual)
- organisms known only from preserved specimens (morphospecies)
- fossils (palaeospecies)
- single specimens
- ring species, because individuals from opposite extremes of the range may not interbreed.

The reproductive isolation of biospecies takes different forms:

- members of different species may have different appearances or behaviour patterns so that courtship does not begin or is not completed
- mating may not be physically possible
- fertilisation may fail, or the zygote may fail to develop
- the species may reproduce at different times or in different places
- interbreeding may occur, but the hybrids are infertile (e.g. mules).

Speciation is the evolution of two or more groups that do not interbreed, and are therefore new species. There are two ways in which one group of interbreeding organisms can produce another group that cannot interbreed with the first.

- **Allopatric** speciation: this requires geographical isolation and time and involves a combination of genetic drift and natural selection. Over time, allele frequencies change and may become so different that, even if the populations are no longer separated, they cannot interbreed successfully. This process is seen in its intermediate stages in many island populations isolated from the same species on the nearest mainland, such as the wren populations on the islands and mainland of Scotland.
- **Sympatric** speciation: this occurs when organisms are not geographically isolated. Speciation can occur in plants in a single generation as a result of **polyploidy**: the different chromosome numbers of parents and offspring provide reproductive isolation. The grass *Spartina anglica* is isolated from its parent species, *S. maritima* and *S. alterniflora*, in this way. Reproductive isolation in space and time may occur even when populations are not geographically isolated. Populations of the fruit fly *Rhagoletes pomonella* mate only on the particular species of fruit tree on which they hatched, each at a different time of the year.

Examiner tip

Remind yourself about classification and phylogeny from Unit 2 of your AS course.

✓Quick check 3

Hint

A ring species is a series of populations, each of which can interbreed with the population on each side, but in which the more distant populations cannot interbreed. Such a series may form a geographical ring.

Examiner tip

n = haploid;
$2n$ = diploid;
$4n$, $8n$ etc. = polyploid.

✓Quick check 4 and 5

✓Quick check 6 and 7

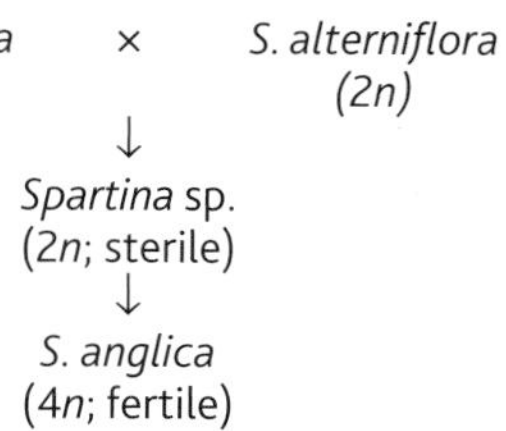

QUICK CHECK QUESTIONS

1. Explain what is meant by the term 'genetic drift'.
2. Distinguish between stabilising selection and directional selection.
3. What is a biospecies?
4. State the difference between allopatric and sympatric speciation.
5. Explain why isolation is important in speciation.
6. Explain how polyploidy provides instant reproductive isolation.
7. What is the likely outcome of the behaviour of the different populations of *Rhagoletes*?

Artificial selection

Key words

- artificial selection
- selective breeding
- polygene
- performance testing
- progeny testing

Artificial selection occurs when a breeder chooses individuals with desirable phenotypes for breeding and/or prevents those with less desirable phenotypes from breeding, thus changing allele frequencies in the population.

Selective breeding

Domesticated animals and crop plants have been subjected to artificial selection for a very long time through **selective breeding**. Individuals showing one or more of the desired traits have been chosen for breeding and, in turn, the most suitable individuals among the offspring have been selected. Over many generations, the alleles giving the desired traits have increased in frequency, while other alleles have decreased in frequency or been lost.

✔*Quick check 1*

Examiner tip

Revise selective breeding from Unit 2 of your AS course.

Artificial selection may be compared with natural selection:

Artificial selection	Natural selection
Selective agent = breeder	Selective agent = total environment of the organism
Traits of use to the breeder are selected, including those that may not be to the organism's advantage	Adaptations to the prevailing conditions are selected
A single trait may be selected	Many different traits contributing to fitness are selected

✔*Quick check 2*

Successful artificial selection depends on choosing parents with desirable traits produced largely by their genotype, not by the environment (see page 56).

Many of the desirable traits of crop plants and livestock are controlled not by single genes, but by many genes: **polygenes** (see page 56). There is often a large environmental component to the variation seen, which is why selective breeding for such characteristics is unpredictable. There is no guarantee that parents with desirable traits will produce the desired offspring.

✔*Quick check 3*

Examples of artificial selection: dairy cows and bread wheat

Unconscious selection of cattle was probably applied by the earliest farmers neglecting the least desirable animals. Methodical selection for traits of economic importance began in the eighteenth century. Selection for fertility, milk yield and milk quality has given rise to the modern dairy cow.

✔*Quick check 4 and 5*

Selection of a suitable female for breeding is done by comparing the performance of different animals under the same conditions (**performance testing**). Selection of males, which do not themselves show the wanted traits, is done by **progeny testing**. The performance of the female offspring of different males is compared. Remember that in cattle breeding, the generation time is long and the number of offspring small, so it takes time to build up a large number of animals with the desired phenotype. It may be necessary to breed from less desirable animals simply to increase numbers. To help with this problem, large numbers of embryos from chosen parents can be produced and transferred into surrogate mothers. Cattle embryos can also be cloned (see page 63).

Bread wheat, *Triticum aestivum*, owes its origin to hybridisation between a number of different, but related, species, as shown in the diagram. Chance doubling of the chromosome number (polyploidy) of the resulting sterile interspecific hybrids restored fertility on at least two occasions. Since the appearance of *T. aestivum*, selection has been for traits such as increased yield, shorter stalks and disease resistance. Selective breeding is quicker in plants than in animals, because a small number of selected plants can produce large numbers of offspring and because self-fertilisation and vegetative propagation (page 62) are often possible.

Hint

'Interspecific' means 'between species'.

✓ Quick check 6 and 7

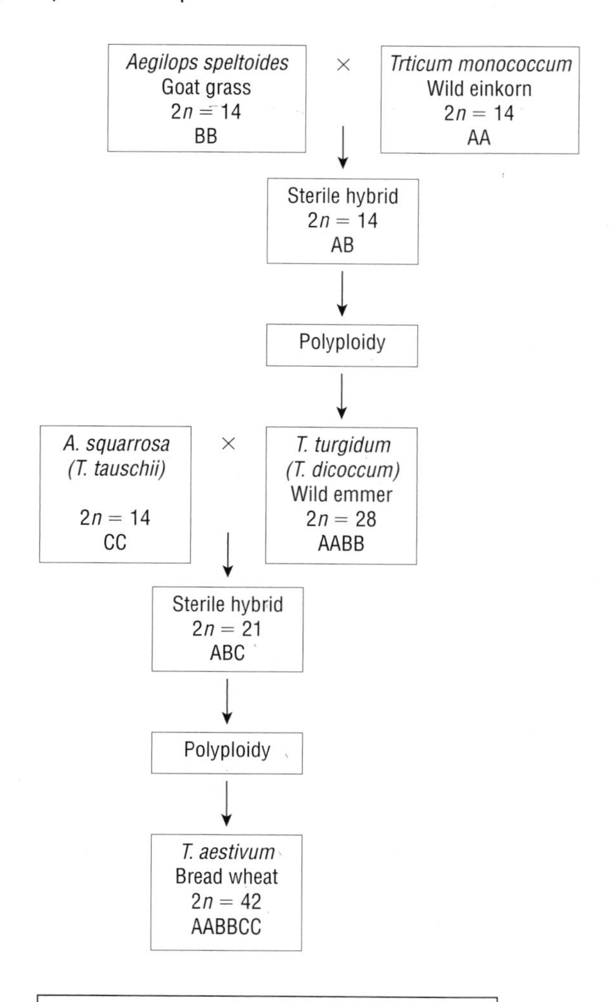

QUICK CHECK QUESTIONS

1 Explain the purpose of artificial selection.

2 Distinguish between natural selection and artificial selection.

3 Many desirable traits in crop plants are controlled by polygenes. Explain why this presents problems in a selective breeding programme.

4 Distinguish between performance testing and progeny testing.

5 Explain why progeny testing is necessary to choose bulls for a selective breeding programme to increase milk yield.

6 Explain why the hybrid resulting from crossing *Aegilops speltoides* and *Triticum monococcum* is sterile.

7 How does polyploidy restore fertility to a sterile interspecific hybrid?

Cloning in plants and animals

Reproductive cloning is the production of a new individual with the genotype of an existing one. **Non-reproductive cloning**, or therapeutic cloning, supplies replacement cells whose genotype matches that of an existing organism.

Key words

- reproductive cloning
- non-reproductive cloning
- vegetative propagation
- sucker
- tissue culture
- micropropagation
- explant
- callus
- surrogate
- somatic nuclear transfer

Module 2

✔Quick check 1

✔Quick check 2

Vegetative propagation

Plant **vegetative propagation** is a form of natural cloning. Many plants have methods of reproducing asexually by separation of some part of the plant body and its subsequent development into a complete plant. For example, English elm trees, *Ulmus procera*, can be propagated by removing **suckers** from a tree in autumn and growing them in a nursery bed. A sucker is a shoot that grows from below ground level, usually from a root. Elms can also be 'layered': low-growing branches can be pegged down onto the soil in autumn, left until they grow roots, and then separated.

Artificial clones of plants can be produced in large numbers by means of **tissue culture** or **micropropagation**.

Tissue culture (micropropagation)

Micropropagation is used for plants that:

- are difficult to grow from seed
- do not have a natural method of asexual reproduction
- are desirable hybrids and do not breed true
- have been genetically engineered.

Hints

An explant is a piece of plant stem, root or leaf that includes some undifferentiated cells (meristematic cells).

Aseptic conditions are those with no contaminating microorganisms.

A callus is a mass of undifferentiated cells.

Orchids are difficult to grow from seed, but are grown in enormous numbers for their flowers using the following technique.

- **Explants** of tissue are taken from young, developing stems that have been surface-sterilised.
- In aseptic conditions, the explants are placed in a growth medium containing sucrose as an energy source, other organic nutrients such as amino acids and vitamins, a range of inorganic ions, and plant hormones to stimulate mitosis.
- Undifferentiated cells in the explant divide by mitosis to produce a **callus**.
- The callus can be subdivided many times to increase the number of plants that can eventually be produced.
- Small pieces of callus are transferred to a new growth medium with plant hormones to trigger differentiation into shoots and roots.
- The resulting embryoids grow into plantlets, which are transferred via a hardening medium into sterile soil.

✔Quick check 3

Some advantages and disadvantages of plant cloning in agriculture are shown in the table.

Advantages	Disadvantages
Very many genetically identical plants can be produced from one original plant	Because the plants are genetically identical, they are all susceptible to a newly mutated pathogen or pest, or to changing environmental conditions
Plants can be produced at any time of the year and air-freighted around the world	The process is labour-intensive – it is more difficult to plant plantlets than to sow seed
Callus can be genetically engineered	

✔Quick check 4

Animal cloning

Artificial clones of animals are produced in two ways.

- By division of an early embryo (16–32 cells). At this stage, all the cells can differentiate into any tissue. It is relatively common in selective breeding of cattle to superovulate a female with the desired traits, fertilise the eggs *in vivo* or *in vitro* with sperm from a suitable male, and then subdivide the resulting embryos. The embryos are implanted in healthy recipient females, which act as **surrogate** mothers. The desirable female is not put at risk in pregnancy and is available for further superovulation. This increases the stock of selectively bred animals.
- By transferring a nucleus from a cell into an egg cell with the nucleus removed (*enucleate* egg). This is **somatic nuclear transfer**. Nuclei from a cell culture of early *embryo* cells were used to produce the first mammals to be cloned successfully by nuclear transfer: the Welsh Mountain sheep Megan and Morag (1995). The sheep Dolly (1997) was the first mammal to be cloned by nuclear transfer from an *adult* cell. A nucleus from a culture of mammary gland cells from a 6-year-old Finn Dorset ewe was put into an enucleate egg from a Scottish Blackface ewe. Dolly was a Finn Dorset.

Hint

Superovulation involves treating a female mammal with hormones so that many egg cells (oocytes) mature in the ovaries at the same time.

✓ *Quick check 5 and 6*

Nuclear transfer is used in the following ways.

- In breeding endangered species. For example, a cell from an adult African Cape buffalo was fused with an enucleate egg cell from a domesticated cow, introducing both a nucleus and some mitochondria of the Cape buffalo. The resulting embryos were implanted into a surrogate.
- To produce transgenic animals for 'pharming' human chemicals. For example, the sheep Polly (1997) carried the gene for human factor IX, which was secreted into her milk.
- To produce embryonic stem cells that are almost genetically identical with the donor of the nucleus.
- In producing largely human cells for research into various diseases. For example, a nucleus from a human patient suffering from, say, Parkinson's disease can be put into an enucleate cow's or rabbit's egg and the resulting 'cybrid' grown for up to 14 days for research purposes.

Hint

in vivo – in life;

in vitro – literally 'in glass' (in a dish).

✓ *Quick check 7*

Examiner tip

Remember that mitochondria have DNA coding for a few genes (about 13).

There are advantages and disadvantages of artificially cloning animals. Advantages include:

- the reproductive rate of a genetically superior animal is increased
- the number of animals with a wanted trait is increased
- the cloned embryo can be sexed or tested for certain genetic diseases before implantation into a surrogate
- a fertile female of an endangered species is not needed for somatic nuclear transfer.

A major disadvantage is that ethical objections can be raised against many of the procedures. The animals concerned are not only being denied their natural instincts and behaviour, but are being used by the breeder as a means to an end.

QUICK CHECK QUESTIONS

1 Distinguish between reproductive and non-reproductive cloning.

2 Describe vegetative propagation in English elm trees.

3 Explain what is meant by 'callus'.

4 State one advantage and one disadvantage of cloning plants in agriculture.

5 What is a surrogate mother?

6 Describe the process of somatic nuclear transfer.

7 What is the genetic content of a human/rabbit 'cybrid'?

Biotechnology

Module 2

Key words

- biotechnology
- primary metabolite
- secondary metabolite
- binary fission
- generation time
- limiting factors
- batch culture
- continuous culture

Quick check 1

Biotechnology is the industrial use of living organisms, or parts of living organisms, to produce food, drugs or other products.

Microorganisms are often used in industrial processes because they:

- are highly versatile, occupying a wide range of habitats, including extreme conditions
- have a rapid growth rate
- are small, and can be produced in large numbers in a small volume
- can be grown in the laboratory, so are not influenced by climate
- produce more enzymes per unit mass than larger organisms
- can be genetically manipulated to express genes for novel products or human biochemicals
- can be manipulated to produce secondary metabolites on demand.

Hint

A **primary metabolite** is the product of an organism's essential metabolism (primary metabolism), e.g. liberating energy and producing biomass.

Secondary metabolites include products used in metabolism that occurs only at specific points in the life cycle, and products that have no apparent use. They are often produced by cells that have stopped dividing. Penicillin is a secondary metabolite.

Quick check 2

Growth of a microorganism in closed culture

Bacteria multiply by **binary fission**, in which one cell divides into two daughter cells, doubling the number of cells every generation. The time taken for a bacterial population to double is called its **generation time**. Under optimal conditions, some bacteria can divide as frequently as once every 20–30 minutes. But there are always factors preventing unlimited population growth, such as:

- depletion of nutrients
- depletion of oxygen
- accumulation of toxic or acidic waste products.

These limiting factors are density-dependent: the greater the number of individuals in the population (density), the greater the effect of the factor.

Growth of a bacterial population in a closed system, such as a culture flask, is shown in the figure.

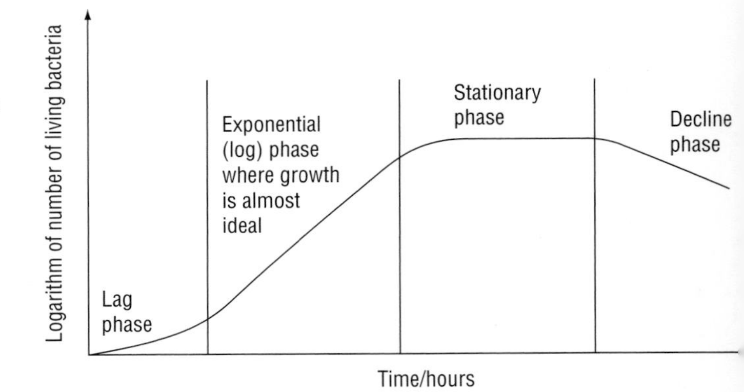

There are four recognisable phases of population growth.

- Lag phase: the bacteria adjust to new conditions, synthesising carriers to absorb nutrients or enzymes to digest them. This may involve switching on genes (see *lac* operon, page 46). Growth is slow.
- Exponential phase (logarithmic phase): the numbers of bacteria double in each unit of time. The exponential (geometric) growth curve that this gives becomes a straight line when the logarithms ($\log_{10}$) of the numbers of bacteria are plotted against time. There are no restrictions to growth in this phase, and growth is rapid.
- Stationary phase: limiting factors, such as the failing supply of nutrients and oxygen and the build-up of waste products, gradually have their effect until the death rate equals the division rate and no population growth occurs.
- Decline phase (death phase): the death rate is greater than the division rate so the population becomes smaller, although the decline may be slowed slightly by breakdown of cells releasing nutrients that other cells can use.

Examiner tip

Compare this use of the concept of limiting factors with that in photosynthesis on page 24.

Examiner tip

Plotting the logarithm ($\log_{10}$) of the number of individuals converts the exponential (logarithmic) curve into a straight line.

Quick check 3

Immobilised enzymes

Enzymes can be immobilised by being:
- adsorbed or bonded onto an insoluble matrix, such as collagen or cellulose
- held inside a gel lattice, such as silica gel
- held inside a partially permeable membrane, such as cellophane
- held in a microcapsule of polyacrylamide or alginate.

The advantages of using immobilised enzymes include:
- the enzyme can be recovered easily and used many times
- the product is not contaminated with the enzyme, making the process ideal for continuous culture
- protection by the immobilising material means the enzyme is more stable in changing temperatures or pH
- enzyme activity can be controlled more accurately.

✓Quick check 4

Batch culture and continuous culture

Biotechnologically useful microorganisms can be grown in **batch culture** or in **continuous culture**. The two procedures are compared in the table.

Batch culture (e.g. production of penicillin)	Continuous culture (e.g. production of mycoprotein)
Carried out in a closed fermenter	Carried out in an open fermenter
Nothing is added; only waste gases removed	Nutrients are added continuously
Product is separated from mixture at end of process	Product is removed continuously
The microorganism's exponential growth phase is short	The microorganisms are kept in the exponential growth phase
Easy to set up and control	More difficult to control than batch culture
The fermenter can be used for different purposes at different times	The fermenter can be smaller than in batch culture for the same yield
Only one batch is lost should the culture become contaminated	Potential losses from contamination are larger than in batch culture because productivity is greater

To maximise the yield of products from a fermenter, the following growing conditions must be manipulated so that they are optimal for the exponential growth phase: temperature, pH, concentration of oxygen (for aerobes), water potential, concentration of nutrients and concentration of waste products.

Asepsis is essential. Unwanted organisms in the fermenter:
- compete with the wanted organism for resources, reducing the yield
- kill the wanted organism (e.g. bacteriophage viruses kill bacteria)
- produce unwanted products, which may be toxic
- may themselves be pathogenic (disease-causing).

✓Quick check 5

QUICK CHECK QUESTIONS

1 Define the term 'biotechnology'.

2 State two factors that could prevent unlimited population growth of a microorganism.

3 Explain why, after a while, growth of a population of bacteria in a closed system ceases to be exponential.

4 State the advantages of using immobilised enzymes.

5 Distinguish between batch culture and continuous culture of microorganisms.

Genomes and gene technology

Module 2

Key words

- nucleotide
- DNA polymerase
- primer
- electrophoresis
- agarose
- Southern blotting

Examiner tip

A DNA **nucleotide** consists of the sugar deoxyribose, a phosphate group and a nitrogenous base – one of adenine (A), thymine (T), cytosine (C) or guanine (G). Review the structure and replication of DNA from Unit 2 of your AS course.

Hint

A DNA nucleo<u>side</u> is the base + sugar part of a nucleo<u>tide</u>. A nucleoside with three phosphate groups attached to it can be called a nucleo<u>side</u> <u>tri</u>phosphate or a nucleo<u>tide</u> <u>di</u>phosphate.

Quick check 1

Quick check 2

Quick check 3

Quick check 4

Sequencing the genome

Sequencing involves breaking DNA into short fragments and determining their base sequences. There are two main methods of sequencing DNA: by *chain termination* or by *chemical cleavage*. Gene sequencing allows genome-wide comparisons to be made between individuals of the same species, and between species, using appropriate databases. Such comparisons increase our understanding of the role of a particular gene, and may suggest a relationship between different species.

A DNA nucleoside triphosphate (NTP) consists of deoxyribose, a nitrogenous base and three phosphate groups. Only NTPs can be used in the replication of DNA. Loss of two of the phosphate groups supplies the energy for the reaction.

In the chain termination method of sequencing DNA, fragments of the DNA to be sequenced are produced by the use of four dideoxynucleoside triphosphates (ddNTPs). A ddNTP lacks a hydroxyl group, which is necessary for a DNA chain to continue growing. Inclusion of a ddNTP terminates a growing DNA chain.

DNA is denatured to separate its two strands. In the presence of an *ample* supply of each of the four NTPs, single-stranded DNA is replicated by **DNA polymerase**, using a **primer** to begin the synthesis. A primer is a short length of single-stranded DNA complementary to the base sequence at the 3′ end of the chain. It forms a starting point from which DNA polymerase can continue replicating the chain.

A *low* concentration of one of the ddNTPs (e.g. the C form) is present so that it will be added only rarely to a lengthening chain:

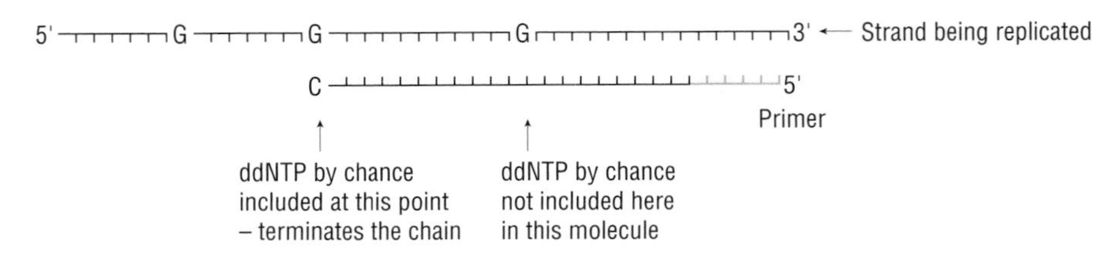

Similarly, in separate reactions, each of the three other ddNTPs (A, T and G) is added to identical strands of DNA.

Adding the products of all four reactions together produces a set of fragments that end at nucleotides with different bases, and that differ in length by one nucleotide. This is called a set of **nested fragments**.

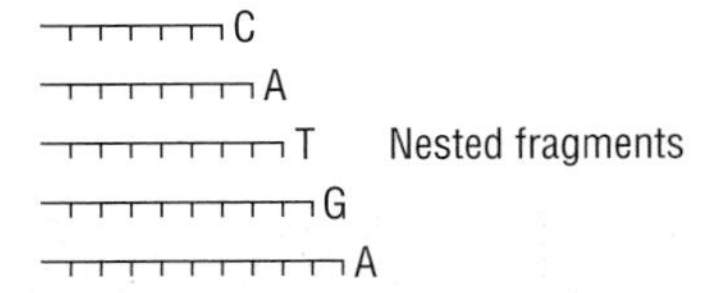

The nested fragments are sorted by length by gel **electrophoresis**.

Automated sequencing can be done by labelling each of the four ddNTPs with a different-coloured fluorescent dye. On separation in a single lane of a sequencing gel, the different colours can be distinguished as the fragments leave the end of the gel and pass through a detector. A computer displays the data as a series of peaks, giving a direct reading of the sequence.

Electrophoresis

DNA fragments can be separated according to length by **agarose** gel electrophoresis, as follows.

- The DNA sample is placed in a well at one end of a gel in an electrophoresis bath filled with buffer (electrolyte).
- A direct current is passed thought the gel.
- DNA fragments are negatively charged and so move towards the anode (positive electrode).
- The shorter the fragments, the further they travel in the time allowed.

✓Quick check 5 and 6

The pattern of bands of DNA is invisible at this stage, unless a blue or fluorescent dye which attaches to DNA is added. Usually the invisible DNA banding pattern is transferred to a nylon membrane by **Southern blotting**.

- The gel is covered with a nylon membrane and then by absorbent paper towels.
- The DNA is drawn up onto the membrane by capillarity.
- The membrane is then heated to denature the DNA into single strands.
- A radioactive (^{32}P) DNA probe with a base sequence complementary to part of the wanted sequence is added, and surplus washed off.
- The position of the bound probe can be found by placing X-ray film over the membrane. The emissions from the radioactive probe produce black bands on the developed film.

✓Quick check 7

QUICK CHECK QUESTIONS

1. What is meant by a dideoxynucleoside triphosphate (ddNTP)?
2. Describe the role of a primer in the replication of DNA.
3. Explain what is meant by a set of nested fragments.
4. What allows gene sequencing to be automated?
5. Why do DNA fragments move to the anode in gel electrophoresis?
6. How does electrophoresis sort fragments of DNA by length?
7. Describe the process of Southern blotting.

Gene technology

Key words
- recombinant DNA (rDNA)
- genetic engineering
- restriction enzyme
- reverse transcriptase
- blunt ends
- sticky ends
- ligase
- polymerase chain reaction (PCR)
- *Taq* polymerase
- vector
- plasmid

Recombinant DNA (rDNA) is DNA that has been made by **genetic engineering**, by joining together pieces of DNA from two or more different sources.

Genetic engineering is the extraction of a gene from one organism, or the manufacture of a gene, in order to place it in another organism in such a way that the receiving organism expresses the gene.

A gene that is wanted for genetic engineering may be obtained in different ways.
- The gene may be identified in the donor organism's DNA and extracted using **restriction enzymes**.
- mRNA transcribed from the wanted gene may be extracted from the donor organism and converted to single-stranded complementary DNA (cDNA) by the enzyme **reverse transcriptase**. The single-stranded DNA is made double-stranded by DNA polymerase (see page 70).
- The gene may be manufactured by using the triplet code, if the primary structure of the wanted protein is known.

Quick check 1 and 2

Quick check 3

Restriction enzymes

Restriction enzymes (restriction endonucleases) are found in bacteria, where they break down the DNA of invading bacteriophage viruses ('phages) and so restrict the viral multiplication.
- Each restriction enzyme binds to and cuts DNA at a specific target site.
- The target site is commonly between 4 and 6 base pairs long, and is symmetrical (palindromic).
- The bacterium's own DNA is protected from attack by not having the target site, or by having it hidden by chemical markers.
- The two strands of DNA may be cut in the same place, leaving **blunt ends**:

  G T T↓A A C → GTT and AAC
  C A A↑T T G → CAA and TTG

  or cut in different places, leaving **sticky ends** of unpaired bases:

  G↓G A T C C → G and GATCC
  C C T A G↑G → CCTAG and G

Hint

Restriction enzymes are identified by an abbreviation indicating their origin. This consists of the first letter of the genus and the first two letters of the species. (There may also be a strain descriptor.) A Roman number is added to distinguish enzymes from the same source. *Hpa*I was the **first** restriction enzyme found in *Haemophilus parainfluenzae*.

Joining DNA to form rDNA requires complementary single-stranded sequences: sticky ends. Hydrogen bonding between complementary bases restores the structure, except for covalent bonds in the sugar–phosphate backbones of DNA. DNA **ligase** restores these bonds.

Quick check 4 and 5

Hint

A ligase is an enzyme catalysing the condensation of two molecules.

The polymerase chain reaction

Multiple copies of a wanted DNA fragment are made by the **polymerase chain reaction (PCR)**. Each cycle of the reaction has three steps:
1. denaturation: DNA is heated to separate its two strands
2. annealing: primers (short lengths of single-stranded DNA complementary to part of the sequence near its ends) bind as the temperature is lowered
3. elongation: the primers are elongated by ***Taq* polymerase**.

The three steps are repeated n times, giving 2^n copies of the original DNA.

Taq polymerase is a heat-stable DNA polymerase from the bacterium *Thermus aquaticus*, from hot springs in the Yellowstone National Park. It is not destroyed by the denaturation step and so does not have to be replaced during each cycle. Its high optimum temperature means that the temperature does not have to be dropped below that of the annealing process, so efficiency is maximised.

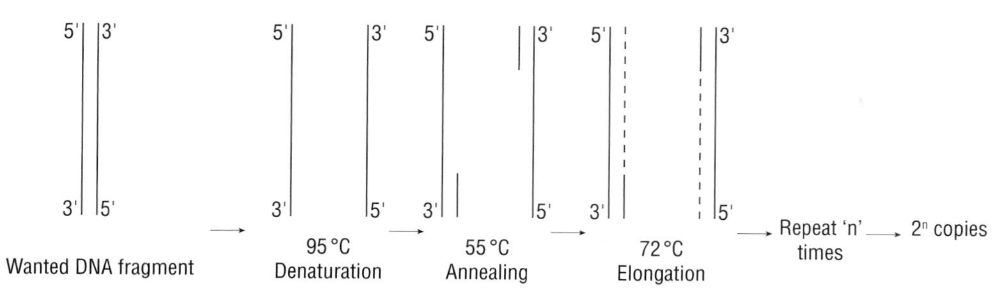

The wanted DNA is inserted into a **vector**, which is put into bacterial or other cells.

Hint

In forensic science, PCR is used to help identify very small samples of tissues (e.g. hairs) as belonging to a specific person.

✓ Quick check 6

Module 2

Vectors

A vector is a means of delivering a gene into a cell. It is a carrier into which a DNA fragment containing the wanted gene can be inserted. The result is recombinant DNA. Commonly used vectors include:

- liposomes
- viral DNA, including bacteriophages ('phages)
- bacterial **plasmids**
- hybrid vectors with the properties of both 'phages and plasmids: cosmids, phagemids and phasmids.

Plasmids

A plasmid is a small, circular DNA molecule found in prokaryotes. It carries genes that are not essential for cell growth or division, but which confer traits that can be a selective advantage under certain conditions.

For genetic engineering, naturally occurring plasmids have been modified to produce vectors that have the desired characteristics. For example, the pUC family of plasmids have all the characteristics of a good vector. They have:

- a low molecular mass, so that they can be taken up readily by bacteria
- an origin of replication, so that they can be copied
- several single target sites for different restriction enzymes in a short length of DNA called the polylinker
- a gene that will be inactivated when foreign DNA is inserted, allowing identification of transformed cells that have taken up the recombinant vector, rather than the original, unchanged plasmid. In this case, the gene (*lac*Z) codes for β-galactosidase. Insertion of a length of DNA into the polylinker region results in a non-functional enzyme. This is the basis of a screening method using the substrate X-gal. X-gal is a colourless substrate that is hydrolysed by β-galactosidase to a blue product. Cells containing recombinant plasmids grow as white colonies on a medium containing X-gal, whereas cells containing 'empty' vectors (non-recombinant plasmids) have functional β-galactosidase and produce blue colonies.

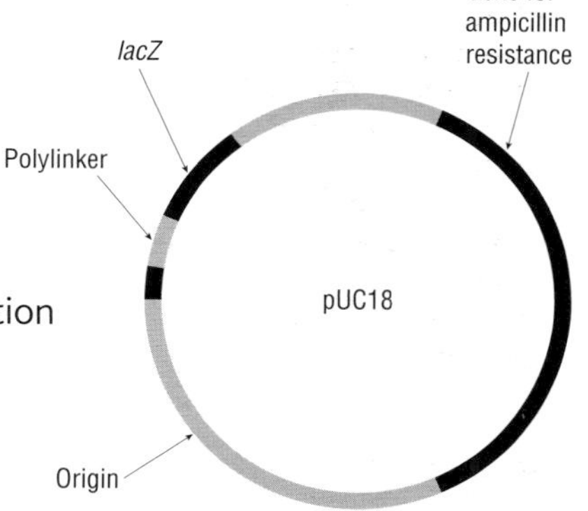

✓ Quick check 7

QUICK CHECK QUESTIONS

1. State what is meant by 'recombinant DNA (rDNA)'.
2. Explain what is meant by genetic engineering.
3. List three ways of obtaining a gene needed for genetic engineering.
4. Explain what is meant by a 'restriction enzyme'.
5. Distinguish between blunt and sticky ends on a piece of DNA.
6. List the three steps of each cycle of the polymerase chain reaction (PCR).
7. What is the role of a vector in genetic engineering?

Genetic engineering

Key words

- transformation
- xenotransplantation
- gene therapy
- somatic cell
- germ line cell

Hint

As DNA in all organisms has the same structure, it is possible to join together pieces of DNA from different sources.

Genetic engineering involves the extraction of a gene from one organism, or the manufacture of a gene, in order to place it into another organism in which it will be expressed (see page 68).

Production of insulin

People with type 1 diabetes (see page 13) cannot make insulin and need daily injections of the hormone. The gene for human insulin was inserted into bacteria, using the following steps, so that large quantities could be manufactured.

1. Preparation of insulin gene

- mRNA for human insulin extracted from pancreas cells.
- Reverse transcriptase uses mRNA as a template to make single-stranded complementary DNA (cDNA) and this is made double-stranded by DNA polymerase.
- A single sequence of nucleotides (GGG) is added to each end of the DNA to make sticky ends.

2. Preparation of a vector to carry the human gene into a bacterium

- Plasmid is cut open with a restriction enzyme (see page 68).
- Cut plasmid has a single sequence of nucleotides (CCC) added to each end to make sticky ends.
- Plasmid and insulin gene are mixed so that sticky ends form base pairs.
- DNA ligase links sugar–phosphate backbones of plasmid and insulin gene.

3. Formation of genetically engineered bacteria

- Plasmids are mixed with bacteria in the presence of calcium ions (to increase porosity).
- Bacteria take up plasmids and multiply to form a clone.
- Genetically engineered bacteria transcribe and translate the human gene to make human insulin.

✓*Quick check 1 and 2*

Hint

The ability of bacteria to take up DNA such as plasmid DNA from their environment is an advantage. Such **transformation** allows useful genes, such as genes for antibiotic resistance, to be shared between bacteria of the same or of different species.

Production of Golden Rice™

Another example of genetic engineering involving plasmids and bacteria is the production of Golden Rice™ to control vitamin A deficiency in parts of the world where rice is a staple food. Vitamin A is present in the outer layers of rice, but in tropical conditions this layer is removed to stop harvested rice from rotting. Vitamin A is *not* present in the remaining endosperm (storage tissue).

- Two genes from daffodil and one from the bacterium *Erwinia uredovora* were inserted into Ti plasmids and taken up by the bacterium *Agrobacterium tumefaciens*. This in turn introduced the genes into rice embryos.
- The resulting rice plants produced seeds with β-carotene in the endosperm, which is yellow, hence 'golden'. Vitamin A is produced in our bodies from β-carotene.

✓*Quick check 3*

Agrobacterium tumefaciens infects wounds in plants and causes the production of tumours. A large plasmid, the Ti plasmid, carries a group of tumour-inducing genes, which are added to one of the host's chromosomes. Genes added to the Ti plasmid will also be added to the plant genome.

✓*Quick check 4*

Gene therapy

There have been several attempts to genetically engineer mammals so that their cells do *not* trigger an immune response, and hence rejection, when transplanted into humans. Pigs are often used, as their organs are a similar size to those of humans. Such transplantation between different species, or **xenotransplantation**, produces a massive immune response, but pigs have been produced that do not express the tissue-typing HLA genes and that have a human gene for a cell surface protein that prevents attack by the human body's defences.

Examiner tip

Review the immune response from Unit 2 of your AS course.

A question about this will expect you to use your knowledge of immunity.

Gene therapy is the treatment of a genetic disorder by altering an individual's genotype.

When the disorder is a recessive condition, for example cystic fibrosis or X-linked severe combined immunodeficiency, the therapy aims to add the normal, dominant allele of the defective gene to the genotype. There are two potential approaches to such therapy:

- **somatic** (body tissue cell) therapy
- **germ line cell** therapy of sperm, egg or fertilised egg.

These are distinguished in the flow diagram:

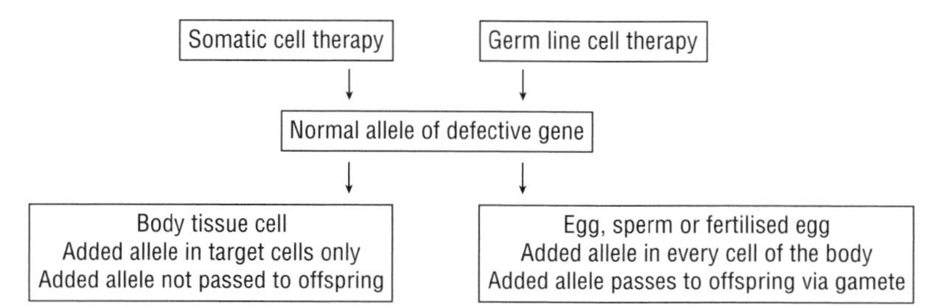

Module 2

✓ *Quick check 5 and 6*

Ethics of genetic engineering

Ethics are sets of standards by which particular groups of people agree to distinguish an acceptable from an unacceptable activity. Ethics change with time, because people alter their views according to their knowledge and experience. Attitudes to genetic engineering in different countries differ widely, and the implications of such techniques are subject to much public debate.

- Ethical objections to genetic engineering range from religious objections to tampering with an organism's natural genotype to fears of unforeseen effects of the gene concerned, or of the consequences of its escape into wild populations. Growing such plants might damage the environment, or eating them might be bad for health.
- Others think it would be unfortunate if public anxieties hampered trials and eventual use of organisms modified to produce much-needed vaccines, antibodies and other pharmaceutical products. Animals engineered to show human diseases allow for valuable research, and gene therapy has the potential to alleviate distressing genetic disorders.

Hint

Human germ line therapy is illegal in the UK.

✓ *Quick check 7*

QUICK CHECK QUESTIONS

1 Describe how the following are used in genetic engineering: reverse transcriptase, restriction enzyme and plasmid.

2 Explain how sticky ends of DNA bond together.

3 State why it is an advantage for bacteria to take up plasmid DNA.

4 State the role of *Agrobacterium tumefaciens* in the production of Golden Rice™.

5 Explain what is meant by gene therapy.

6 Distinguish between somatic cell and germ line cell therapy.

7 Why do ethics change with time?

Ecosystems

Key words

- community
- habitat
- niche
- population
- ecosystem
- biotic
- abiotic
- producer
- primary consumer
- secondary consumer
- food chain
- food web
- decomposer
- bomb calorimeter

Examiner tip

Make sure you can define these terms and use them correctly when writing about ecosystems.

✓Quick check 1 and 2

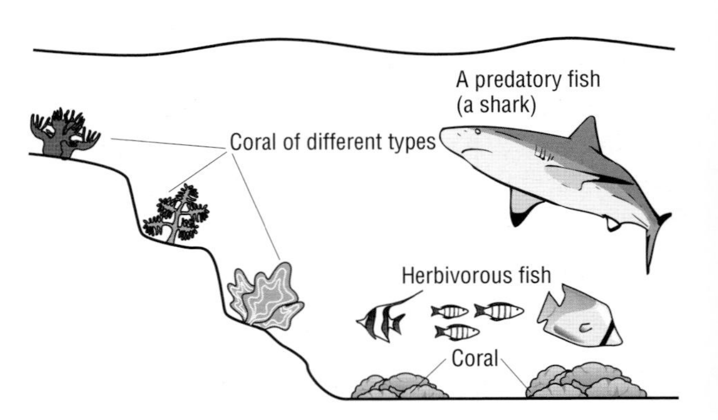

This picture shows part of a coral reef with herbivorous fish grazing on algae that grow over the surface of the rock.

Many organisms live on or around coral reefs, forming a **community**. The reef is their **habitat**. Individual species, like the fish, occupy a particular **niche** on the reef. All the fish of the same species on this reef form a **population**. The reef, with its community of organisms, and the physical environment form a dynamic **ecosystem**, in which the organisms (**biotic** factors) are affected by one another and by factors such as light intensity and temperature (**abiotic** factors).

Energy flows through ecosystems. **Producers**, such as algal films on the reef, convert light energy from the sun into chemical energy. **Primary consumers**, such as herbivorous fish, feed on the algae, becoming food for **secondary consumers**, such as predatory fish. This simple feeding relationship is a **food chain**:

algae → herbivorous fish → predatory fish

In ecosystems, such as coral reefs, the feeding relationships are complex as each animal eats a variety of different foods and is eaten by a number of different predators. **Food webs** show these different feeding relationships.

Ecosystem definitions

Community	All the organisms living in one easily defined area.
Habitat	The place where an organism lives.
Population	All the organisms belonging to the same species living in the same area at the same time. Males and females in a population can interbreed.
Niche	The role of a species in a community. The role refers to its position in a food chain and how it interacts with the environment and with other species.
Ecosystem	A self-contained community together with all the physical features that influence the community, and the interactions between organisms and their environment.
Producer	An organism that converts simple inorganic compounds (e.g. carbon dioxide, water and ions) into complex organic compounds. Most use light to provide the energy to drive the reactions involved.
Consumer	An organism that gains energy from complex organic matter (e.g. herbivores, carnivores, omnivores, detritivores, decomposers, parasites).
Decomposer	An organism that feeds on waste from other organisms, or on dead organisms.
Food chain	Shows energy flow from one organism to another in a simple sequence.
Food web	Shows energy flow through many species in an ecosystem.
Trophic level	Each feeding level in a food chain.

Examiner tip

Revise autotrophs and heterotrophs from Unit 2 of your AS course. See also page 18.

✓Quick check 3

Energy flow

Only a small part of the energy entering a trophic level becomes available to the next trophic level. The percentage varies according to the food chain concerned and the efficiency with which energy is transferred, but is rarely greater than 10%. Not much energy reaches animals at the top of food chains (e.g. tigers). This explains why they are rare.

Sunlight

Producers e.g. algae

Death and decay

Feeding

Respiration

Primary consumers e.g. herbivorous fish

Death and decay

Decomposers

Predation

Secondary consumers e.g. predatory fish

Death and decay

Energy is lost in food chains because animals:

- never eat all the available food
- cannot digest all the food they eat
- use energy in their respiration so they can move, hunt, chew, reproduce, etc.
- lose heat energy to their surroundings
- lose energy in urine and faeces: this energy may pass to decomposers.

Energy transfers can be calculated by measuring the energy content of samples of organisms from each trophic level. Each sample is dried to constant mass in an oven and then burned in oxygen in a **bomb calorimeter** (see diagram). The heat energy produced by the oxidation passes to a known mass of water, and the temperature rise of the water is measured. Given that 4.12 J of heat energy raises the temperature of 1 g of water by 1 °C, the energy content of each sample can be calculated.

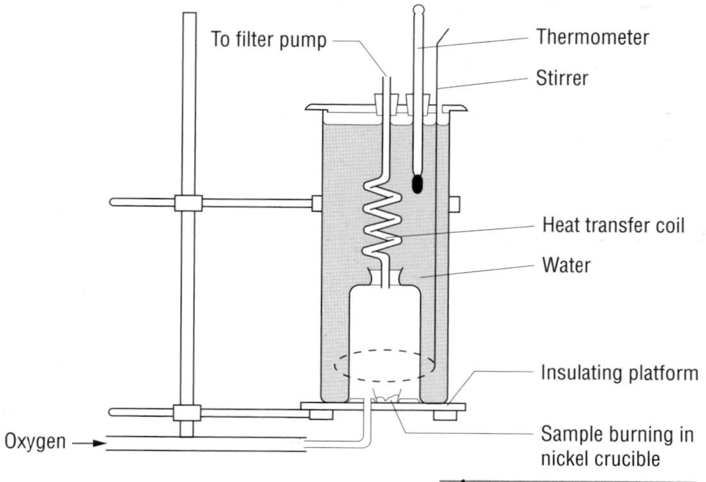

Quick check 4 and 5

Human activities in farming, forestry and fishing manipulate the flow of energy through an ecosystem by altering the productivity of one or more trophic levels. Examples include:

- replacing natural vegetation and fauna with crops and livestock
- deflecting natural succession (see page 74) to maintain grassland
- increasing productivity of producers through soil improvement, irrigation, fertilisers, and removal of competing weeds and damaging pathogens and pests
- increasing productivity of producers and consumers through selective breeding (see page 60) or genetic engineering (see page 70)
- sheltering organisms from damaging environmental factors.

Quick check 6

QUICK CHECK QUESTIONS

1 Distinguish between a producer and a consumer.

2 What do secondary consumers eat?

3 Distinguish between a habitat and a niche.

4 List the ways in which energy may be lost from a food chain.

5 How is the energy content of organisms in a trophic level determined?

6 Explain how human activities can manipulate the flow of energy through an ecosystem.

Succession

Key words

- primary succession
- climax community
- pioneer species
- quadrat
- point frame (quadrat)
- line transect
- belt transect

Succession is a change in the structure and species composition of a community. It occurs because of the changes caused by the presence of established organisms, or by some external influence. The immediate changes caused by agricultural activities are not included in this definition.

The changes that result when the starting point is bare, uncolonised land are called **primary succession**. This occurs in:

- sand dunes
- lava and ash from volcanoes
- landslides
- land and lakes left by retreating glaciers.

The first colonisers on bare rock are lichens and mosses. Their presence allows some soil to build up, allowing ferns and flowering (vascular) plants to appear. When the bare ground is particulate (e.g. made up of sand or ash particles), colonisation is faster because vascular plants can take root without the lichen and moss stage. Eventually the soil becomes deep enough to support shrubs, and then trees. In Europe, woodland is the endpoint of succession on land, and is called the **climax community**.

✔Quick check 1

The endpoint of primary succession in a freshwater pond or lake is also woodland. Once the water has been colonised by plants, there is an increase in organic matter and the lake gradually fills in.

Hint

A *sere* is a particular example of plant succession. A *hydrosere* is a sere beginning in water.

The effects of named organisms on the process of succession can be studied with examples of primary succession, such as sand dune, lake hydrosere or volcanic island. Some of these organisms will be early arrivals (**pioneer species**) and others characteristic of the **climax community**.

Pioneer species:

- are able to tolerate extreme conditions, e.g. low nutrient levels
- have very good means of dispersal, usually by wind
- are not able to compete for resources, e.g. light
- are not influenced by or dependent on animal species
- may be able to fix nitrogen (e.g. legumes) and build up soil nutrients.

Examiner tip

List the possible effects on succession of the species you found when sampling a habitat in your AS course.

Climax community species:

- have large seeds (with a large energy store) so that seedlings can survive low light intensity
- have a specialised niche, e.g. as an epiphyte
- are unable to tolerate great fluctuations in the water content of soil
- are strongly influenced by other organisms, e.g. competitors, herbivores, pollinators, seed-dispersal agents and soil microorganisms.

✔Quick check 2

✔Quick check 3

When an established community is destroyed (e.g. by fire, flood or human activity) a new community develops. Succession takes place in this new community and is referred to as *secondary succession*.

Sampling a habitat

When studying a habitat, you need to know:

- what species are present in the habitat
- the abundance of each species present
- the distribution of each species.

Initially, organisms must be identified, usually by using an identification key or consulting an expert in the field.

Examiner tip

Revise the techniques for sampling a habitat from Unit 2 of your AS course.

Techniques for sampling a habitat

Organisms are distributed unevenly in most environments, so random sampling is used to determine the number and abundance of species present. Random sampling also eliminates any bias on the part of the person taking the samples. The area to be sampled is mapped and given a grid of numbered squares of a size appropriate to the sampling technique. The square to be sampled is determined by use of random number tables, or random numbers generated by a calculator or computer.

✓*Quick check 4*

Module 3

- **Quadrats** are square frames, of a size appropriate for the area to be surveyed, within which you can survey what species are present and estimate the abundance of each by:
  - finding the *percentage cover* – the proportion of the quadrat's area occupied by the species
  - counting the *number of individuals* present and finding the average number per quadrat, or *species density*
  - finding the *species frequency* – the proportion of quadrats with a particular species in them
  - using a subjective scale such as ACFOR (abundant, common, frequent, occasional and rare) or DAFOR (dominant, abundant, frequent, occasional and rare).
  Quadrats are suitable for plants and sessile animals in areas with relatively uniform conditions, where the quadrats must be distributed randomly.
- **Point quadrats** are frames through which long pins (usually 10) are lowered vertically. Each species that touches a pin is recorded, together with the total number of times it is touched. The point quadrat must be positioned randomly.
- **Line transects** are lines across a habitat. All the species touching a line are identified and the position of each species that touches the line is recorded. Transects are suitable for habitats with gradations in conditions. The starting point of the transect should be random, but the line should be placed across the gradation concerned, for example, at right angles to the line of a ditch, stream or hedge; or from the sea to the high tide zone.
- **Belt transects** are quadrats placed sequentially along a line transect, making the transect wider. The quadrats are used as above.

Hint

Using a subjective scale is quick and avoids counting. Accuracy improves with practice.

✓*Quick check 5 and 6*

QUICK CHECK QUESTIONS

1. Explain why colonisation is faster on particulate bare ground than on rock.
2. What adaptations would you expect a pioneer plant species to show?
3. Describe typical features of plant species found in a climax community.
4. Explain when it is necessary to use random sampling techniques.
5. State the advantage of using a subjective scale, such as ACFOR.
6. Choose an appropriate sampling technique for finding the plant species present in: (i) a large area of grassland; (ii) a river bank.

Decomposers, the nitrogen cycle and carrying capacity

Key words

- decomposer
- nitrogen fixation
- nitrogen cycle
- carrying capacity
- density-dependent

Quick check 1

Decomposers are organisms that feed on waste from other organisms, or on dead organisms. They include many fungi, some animals and some bacteria. Decomposers recycle materials (see page 18), including carbon and nitrogen.

Nitrogen cycle

All organisms need nitrogen, as it is a component of essential compounds including amino acids, proteins, nucleotides and nucleic acids (DNA and RNA). Nitrogen in the atmosphere (dinitrogen, N_2) is inert, and few organisms can gain their nitrogen from this source.

Some bacteria, such as *Rhizobium*, are able to fix nitrogen from the air by reducing it to ammonium ions (NH^+_4). This process of **nitrogen fixation** requires much energy. Most nitrogen fixation occurs in swollen nodules on the roots of leguminous plants such as peas and beans. The plants provide sugars, from photosynthesis, to the bacteria so that they have the energy to split the nitrogen molecule (N_2) and combine it with hydrogen. *Rhizobium* and the host plants use these ammonium ions to make amino acids. Both then use amino acids to make proteins.

Nitrogen combined with another element, such as hydrogen, is called fixed nitrogen. Most organisms use forms of fixed nitrogen rather than using dinitrogen. The diagram shows the main transformations that occur to nitrogen in plants, animals and microorganisms.

Examiner tip

There is more to the **nitrogen cycle** than this, but you need to know the roles of these bacteria in cycling nitrogen.

Dinitrogen in air (N_2)

Nitrogen fixation by *Rhizobium*

Denitrification

Nitrification

NH_4^+ → NO_2^- → NO_3^-

Oxidation by *Nitrosomonas*

Oxidation by *Nitrobacter*

Absorbed by plants

Death and decay

Nitrogen-containing molecules in organisms, e.g. amino acids, proteins, nucleic acids

Quick check 2 and 3

Module 3

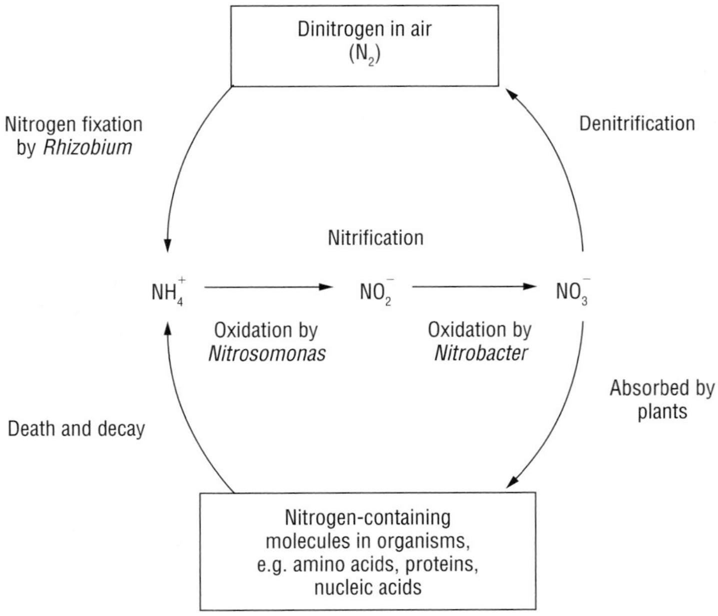

Limiting factors in population growth

Any factor that prevents a population from increasing in size is a limiting factor. At any given time, there is usually just one factor limiting population growth, but different limiting factors come into play at different times. The limiting factor may operate seasonally. Examples of limiting factors for populations of terrestrial flowering plants and animals are shown in the table.

Flowering plants	Animals
Lack of light	Lack of food/prey
Lack of water	Lack of water
Lack of carbon dioxide	Lack of oxygen (likely only in aquatic habitats)
Lack of nutrient ions	Lack of suitable site for reproduction/egg-laying
Inappropriate temperature	Inappropriate temperature
Infection by pathogens	Infection by pathogens
Agent for cross-pollination absent	Lack of a mate at very low population density
Lack of space (another factor almost always becomes limiting before physical space)	Lack of space (many animals defend a territory, ensuring resources for reproduction)

Examiner tip

Compare this use of the concept of limiting factors with that in photosynthesis on page 24.

Module 3

Carrying capacity

The maximum population density of an organism that can be supported permanently in a habitat is called the **carrying capacity**, and is determined by one or more of the **density-dependent** limiting factors. Such factors may be:

- **biotic** – caused by other living organisms (e.g. predation, competition or infection by pathogens)
- **abiotic** – involving non-living components of the environment (e.g. water supply or nutrients in soil).

Hint

A density-*dependent* factor has a proportional increase or decrease in its effect as the population density rises or falls. The effect of a density-*independent* factor is not related to population size.

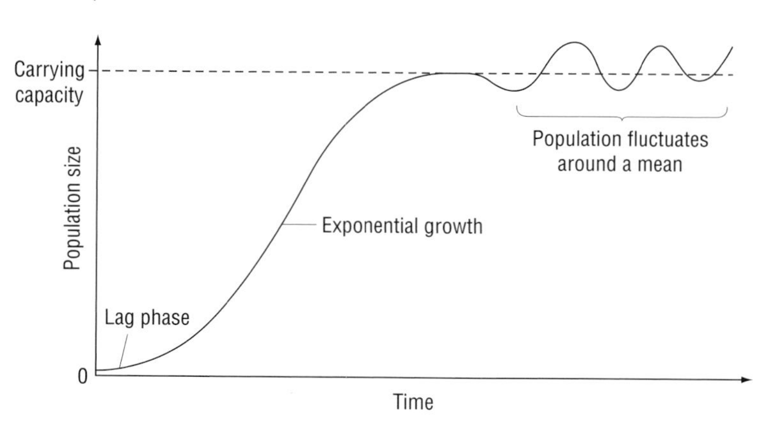

Quick check 4, 5 and 6

QUICK CHECK QUESTIONS

1 State the role of a decomposer in an ecosystem.

2 Explain how nitrogen in the protein of a dead animal is made available to plants.

3 Describe the roles of *Rhizobium*, *Nitrosomonas* and *Nitrobacter* in the nitrogen cycle.

4 Distinguish between biotic and abiotic factors.

5 Explain what is meant by a density-dependent limiting factor.

6 Define the term 'carrying capacity'.

Predator–prey interactions

Key words

- intraspecific competition
- interspecific competition
- competitive exclusion

Examiner tip

In a graph such as this, look for *trends* and *times* of peaks.

Module 3

Annual records of the numbers of Canadian lynx and snowshoe hares trapped for the Hudson's Bay Company in Canada have been kept since the 1820s, and provide evidence of a 10-year cycle in the numbers of both animals (see figure). Both lynx and hare were trapped for their fur. Trappers found that at a peak of the cycle, thousands of lynx could be trapped in a season, while in a trough 5 years later, only a few hundred were trapped.

A possible explanation of this cycle is that:

- a small population of snowshoe hares cannot support many lynx, so the lynx population crashes
- but now there are few lynx to prey on the snowshoe hares, so the hare population starts to rise
- this, in turn, supports more lynx, so the lynx population rises
- but the large numbers of lynx that are eating snowshoe hares causes the hare population to crash – so the cycle begins again.

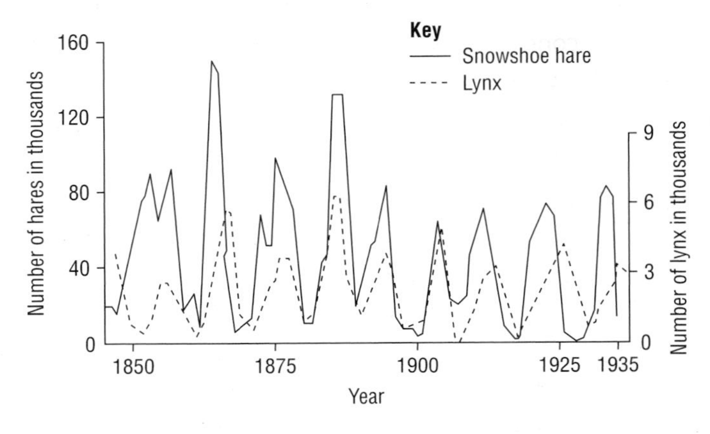

Interestingly, on islands off the east coast of Canada where there are snowshoe hares but no lynx, the hares show the same 10-year cycle of changing numbers as the mainland hares.

This suggests that fluctuations in the number of hares affect the lynx population, but that the numbers of lynx do not cause the fluctuations in the hare population. Instead, this may result from:

- changes in the availability of food plants
- changes in the nutritive quality of food plants
- reduction in breeding success of hares at high population densities.

Other small herbivores, for example lemmings, also show regular cycles in their numbers with no obvious involvement of predators.

✓*Quick check 1 and 2*

Competition

Organisms compete for resources or space that are in limited supply. Such competition falls into two categories:

- **intraspecific competition** – between members of the *same* species
- **interspecific competition** – between members of *different* species.

Hint

intra- = within;
inter- = between.

✓Quick check 3

Intraspecific competition

Density-dependent factors (see page 77) limit population growth, and the individuals with adaptations that are better suited to the prevailing conditions,will outcompete the less well adapted. This is one of the causes of natural selection (see page 58).

Interspecific competition

Competition from other species restricts the spread of any one species into areas where it is less well adapted. The principle of **competitive exclusion** states that, as a result of competition, two species will not occupy the same niche. One displaces the other, so that each adopts a certain way of life in which it has an advantage.

Hint

A niche is a way of life; a habitat is where an organism lives.

- When the protoctists *Paramecium caudatum* and *P. aurelia* are grown together in laboratory culture, in conditions in which both species survive when grown alone, *P. caudatum* always becomes extinct.
- Duckweed, *Lemna minor*, floating on the surface of a tank of water outcompetes its relative *L. trisulca*, which grows submerged in water.
- When two species of flour beetle, *Tribolium confusum* and *T. castaneum*, are placed together in conditions in which both thrive when cultured separately, only one species survives. At lower temperatures and humidity, *T. confusum* outcompetes *T. castaneum*. At higher temperatures and humidity, the reverse occurs.
- In natural UK habitats, the red squirrel is probably in competition with the introduced, larger, grey squirrel. Red squirrels outcompete grey in conifer woods, but are outcompeted by grey squirrels in woods with fewer than 75% conifers.
- When nitrogenous fertilisers are used on a grassland community, some species benefit more than others. Clovers and other legumes benefit least and grasses outcompete them.

✓Quick check 4, 5 and 6

QUICK CHECK QUESTIONS

1 What evidence does the figure on page 78 provide that the snowshoe hare population controls the lynx population, not *vice versa*?

2 The numbers of bank voles and wood mice in a UK wood show seasonal fluctuations in numbers, peaking in autumn and early winter. Both species are eaten by tawny owls. Predict the effect of fluctuations in the prey species on owl numbers.

3 Distinguish between intraspecific and interspecific competition.

4 Explain what is meant by competitive exclusion.

5 Suggest explanations for the results of competition between species of *Paramecium* and between species of *Lemna*.

6 Suggest why nitrogenous fertiliser does not benefit clovers.

Conservation and management

Key words
- conservation
- preservation
- sustainable management

Examiner tip

Review the conservation of biodiversity and of endangered species from Unit 2 of your AS course.

Module 3

Conservation of a species, habitat or ecosystem may be needed because it is at risk from human influence or other pressures. It is always a dynamic process, because it must respond to change. It may be for the benefit of people or other organisms.

Conservation measure	Example
Protection of species	Laws against hunting the blue whale
Protection of habitats	Laws preventing discharge of wastes into rivers
Restoration or reclamation	Erecting sea defences, preventing sand being blown off dunes
Creating new habitats	Digging ponds, planting new woodland
Captive breeding of endangered species	Breeding programmes for the Hawaiian goose
Prevention of succession	Grazing and/or burning of heath or grassland
National action	Laws preventing destruction of bats and their roosts
International action	Convention on International Trade in Endangered Species (CITES) treaty preventing trade in products such as rhino horn

Conservation is not the same as preservation. **Preservation** protects species and/or habitats, for example by creating a nature reserve, while conservation includes the active management needed to maintain or increase biodiversity. Conservation may be needed to counter the effects of modern farming methods.

✓Quick check 1 and 2

Many ecosystems provide economically important resources. For example, aquatic and marine ecosystems yield fish; natural grasslands yield meat and other animal products. **Sustainable management** allows the same area to be exploited indefinitely:

- it does not result in loss of fertility (e.g. by soil erosion)
- the population that is exploited does not become extinct or decline seriously
- biodiversity is maintained – the destruction of other species that share the habitat with the exploited species is avoided.

Unsustainable management results in damage to the habitat and/or depletion of the exploited species, to a point where it no longer provides an economic return, as has happened to cod over much of the North Sea.

✓Quick check 3

Examiner tip

Much easily available information involves tropical rain forest. Although some of this is relevant, OCR has specified timber production in a *temperate* country.

Sustainable forestry

Timber has great economic value and is used in construction, as fuel, and in paper manufacture. Some timber is grown on a small scale on agricultural land, for example as part of hedgerows, but most is from woodlands and forests. These may have developed naturally, or may be plantations.

There are different ways in which forests or woodlands can be exploited sustainably.

✓Quick check 4

- *Selective felling*: some mature trees, diseased trees and unwanted species are harvested, leaving other trees to develop and distribute seeds to fill the gaps. This is more expensive than cutting all the trees on the site (clear felling).
- *Strip felling*: small patches or strips of forest are cleared completely, leaving other patches untouched to cut many years later, after the first areas have regrown. Large areas are not felled at the same time, so loss of species and soil erosion are avoided.

- *Coppicing*: trees are cut down, leaving stumps from which new shoots develop. These grow rapidly because they have a well developed root system. After a few years, the shoots are cut and yield poles, but not large logs. Coppicing can be repeated indefinitely. Small strips or patches are cut in different years, providing a variety of habitats and so producing high biodiversity. There is renewed interest in the method because willow can produce very large masses of wood in a few years. The wood is unfit for construction work, but excellent for paper manufacture or to burn in power stations to generate electricity.

Hint

Only some species of tree can be coppiced. Pines and firs do not develop new shoots from cut stumps. Hazel and sweet chestnut can be used.

The Galapagos Islands

The effects of human activity on plant and animal populations, and the conflicts that can arise between conservation and exploitation, can be seen on the Galapagos Islands.

- In 1934, some islands were set aside as wild life sanctuaries. In 1959, 97% of the land was made a national park, and in 1964, the Charles Darwin Research Station began operating as an international non-government organisation, aiming to protect and conserve the islands and encourage education and scientific research.
- In 1986, the Galapagos Marine Reserve was created.
- Four of the islands are inhabited; the number of visitors to some of the other islands is strictly limited.
- Tourism is important to Ecuador's economy. Organised tourism began in the 1960s and increased rapidly. Wildlife is threatened both by tourism and by the colonisation that accompanies it.
- Early colonists introduced goats, pigs, rats and dogs, which either compete with the native fauna, or eat bird and reptile eggs. Land iguanas, formerly found on all islands, have been lost from some. On larger islands, the goat and pig populations can be controlled but not easily eliminated.
- Thousands of Galapagos tortoises were killed by whalers and sealers in the eighteenth and nineteenth centuries. Some were taken from one island and abandoned on another, mixing up genomes.
- The breeding project at the Charles Darwin Research Station is well established and helped by recent genotype testing.
- Measures are in place to avoid accidental introduction of an organism to the islands, or accidental transfer from one island to another.
- Some islanders see the national park as a barrier to making a living, and resort to drastic measures – arson on the island Isabela destroyed 10 000 hectares in 1994, and in 1995 armed fishermen occupied the Charles Darwin Research Station during a protest against fishing restrictions.

✔ Quick check 5, 6 and 7

QUICK CHECK QUESTIONS

1. Explain the difference between conservation and preservation of species.
2. List three examples of conservation measures.
3. List ways in which sustainable and unsustainable management of an ecosystem differ.
4. Explain the difference between selective felling and clear felling of woodland.
5. What are the aims of the Galapagos national park?
6. List three major problems for Galapagos wildlife.
7. Explain why there is a conflict between tourism and conservation in the Galapagos.

Plant responses

Key words

- tropism
- phototropism
- coleoptile
- auxin
- apical dominance
- gibberellin
- gibberellic acid (GA)
- abscisic acid (ABA)
- ethene
- abscission

Hint

Auxin, gibberellic acid, cytokinin, abscisic acid and ethene are examples of plant hormones.

Like mammals, flowering plants need a communication system in order to respond adaptively to changes in their internal and external environments. A seed must germinate in suitable conditions, and the radicle and plumule must grow downwards and upwards, respectively. The plant must then grow adaptively with respect to light intensity and variations in water content of the soil. A plant growing in a region with seasons must respond to those seasons and flower at the right time. The plant may need to respond to predators or abiotic stress.

The communication system of flowering plants involves:

sensor (receptor) → plant hormone → effector

Tropisms

✓Quick check 1

Many plant responses are by means of growth. A growth response is called a **tropism** when the growth is in a direction determined by a stimulus. A response towards the stimulus is said to be positive, and one away from the stimulus negative. The table shows some tropisms.

✓Quick check 2 and 3

Stimulus	Growth response
Light	Phototropism
Gravity	Geotropism
Touch	Thigmotropism

Hint

A coleoptile is a protective sheath surrounding the first leaf of a germinating grass or cereal.

If a flowering plant is to respond to a change in its internal or external environment, it must have a receptor that is sensitive to that particular change. For example, a plant that can respond to light by growing towards the source of light (positive **phototropism**) needs a photoreceptor molecule that is affected by light. Much work has been done on the phototropic response of cereal **coleoptiles**, in which the tip of the coleoptile is the receptor region of the plant. A cereal coleoptile responds most to wavelengths in the blue region of the spectrum around 450 nm. It is not yet certain what the photoreceptor is, but two molecules in the coleoptile tip have absorption spectra with maxima in that region: carotene and riboflavin. It is likely that riboflavin is the photoreceptor for phototropism.

The plant's effector is the region of cell elongation, just behind a shoot, coleoptile or root tip. In positive phototropism of a coleoptile, the cells on the shaded side elongate more than those on the illuminated side. The coleoptile then bends towards the source of light. **Auxin** produced by the dividing cells of the tip is transported to cells on the shaded side. It travels down that side, allowing the shaded cells to take up more water. Because of the construction of their cell walls, the resulting expansion is not in their width, but in their length.

✓Quick check 4 and 5

Auxins and apical dominance

In many plants, side shoots do not grow if the main shoot is growing. However, if the bud at the tip of the main shoot (the *terminal* or *apical bud*) is removed, then the *axillary* or *lateral buds* in the axils of leaves lower down the plant will start growing. This effect may be caused by auxins produced by the apical bud inhibiting growth of lateral buds. Removing the apical bud removes the source of the inhibitory auxin. The experimental evidence for this effect of auxin is contradictory, and cytokinins and abscisic acid may also be involved.

As the plant grows taller, the lateral buds furthest away from the tip may start growing. Because the auxin is diluted as it passes down the stem from the apical bud, lateral buds at the base of the stem may no longer be inhibited, because the concentration of auxin reaching them is below the threshold value for inhibition. Once active, the lateral buds in their turn produce auxins and cytokinins. **Apical dominance** gives a plant a shape that allows all parts access to light.

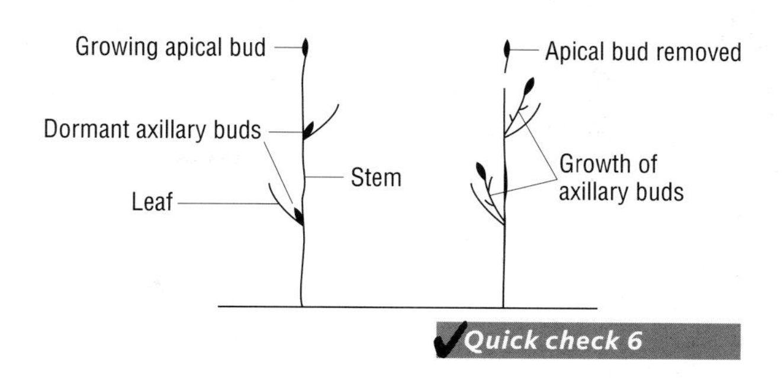

Quick check 6

Gibberellins and stem elongation

Most plants produce **gibberellins** to regulate stem elongation. The effect is seen most obviously when genetically dwarf plants are treated with **gibberellic acid (GA)**: the stems elongate considerably. Similarly, plants with rosettes of leaves separated by very small internodes elongate the internodes when treated with GA. This is seen naturally when a compact lettuce plant bolts (elongates and then flowers).

Abscisic acid and leaf fall

Abscisic acid (ABA) is a plant hormone that stimulates the production of **ethene**. Ethene stimulates the fall of leaves or **abscission**. The following sequence of events leads to leaf fall:

- cells in a layer at the base of the petiole separate from one another by the breakdown of the middle lamellae of their cell walls
- a protective layer of cells with wax in their walls forms on each side of this **abscission layer** to prevent infection and water loss; in woody species this layer is cork with suberised cell walls
- the vascular tissue is sealed
- the leaf is broken off by a mechanical force, such as wind.

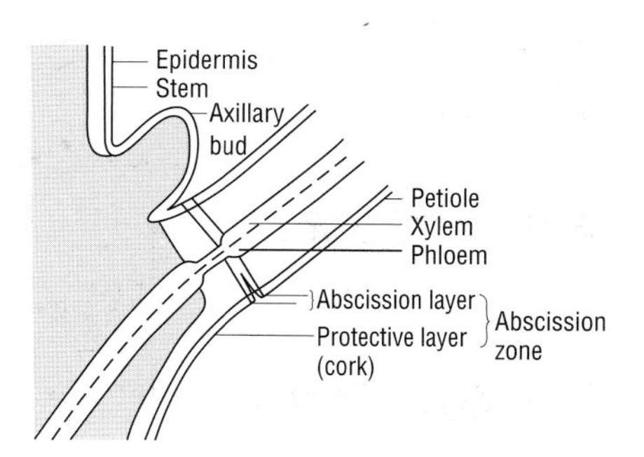

Quick check 7

Commercial use of plant hormones

Farmers and gardeners make use of the effects of some plant hormones. Examples include:

- synthetic auxins used as growth stimulants when rooting cuttings
- a synthetic auxin, 2,4-D (2,4-dichlorophenoxyacetic acid), used as a selective weedkiller – at appropriate concentrations it kills broadleaved species (dicotyledons), but not grass or cereal crops (moncotyledons)
- a form of abscisic acid that is not readily broken down by plants, used as an anti-transpirant as it closes stomata.

Quick check 8

QUICK CHECK QUESTIONS

1. State the essential components of a communication system in a flowering plant.
2. Explain what is meant by a 'tropism'.
3. Name three types of plant hormone.
4. State one example of a receptor in a flowering plant.
5. State one example of an effector in a flowering plant.
6. Explain what is meant by apical dominance.
7. Describe the events leading to leaf abscission.
8. Describe one commercial use of a plant hormone.

Animal responses

Key words
- central nervous system (CNS)
- peripheral nervous system
- somatic nervous system
- autonomic nervous system
- association area
- antagonistic muscles
- flexor
- extensor

All living organisms must respond to their environment. They need a communication system between receptors and effectors, so that they can respond adaptively to changes in both external and internal environments. In animals, responding to changes in the environment is a complex and continuous process, involving nervous (see page 4), hormonal (see page 10) and muscular coordination.

Organisation of the nervous system

The nervous system may be divided into the **central nervous system** and the **peripheral nervous system**.

- The central nervous system (CNS) is made up of the brain and spinal cord.
- The peripheral nervous system consists of the nerves which run between the CNS and the rest of the body – *from* receptors and *to* effectors.

The peripheral nervous system has two components:

- the **somatic nervous system**, which includes all sensory neurones and also the motor neurones that run to skeletal muscles
- the **autonomic nervous system**, which consists of two sets of motor neurones carrying impulses to effectors other than the skeletal muscles, such as glands and the muscles of the gut and heart. Neurones of the *sympathetic nervous system* and *parasympathetic nervous system* use different neurotransmitters and so have different, often antagonistic, effects.

Module 4

Hint

In the parasympathetic nervous system, the neurotransmitter is *always* ACh (see page 8), which often has an inhibitory effect. In *some* sympathetic neurones, the neurotransmitter is noradrenaline, which is stimulatory. Noradrenaline is very similar to adrenaline (see page 10).

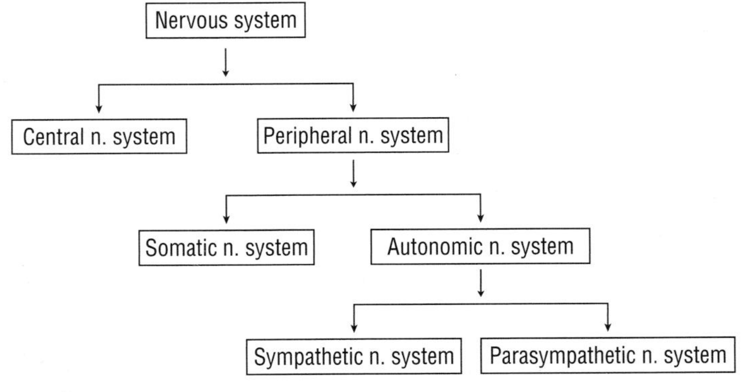

Quick check 1, 2 and 3

The human brain

The annotated longitudinal section of the brain shows its structure and some of its functions.

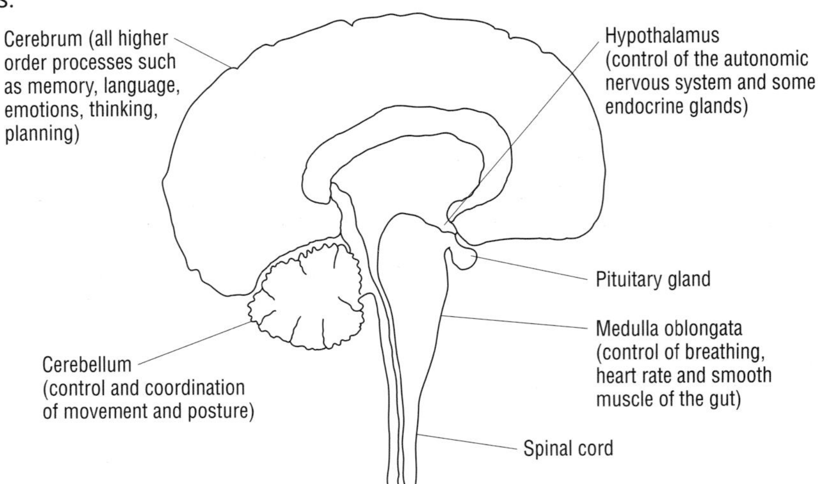

Quick check 4 and 5

Coordination of movement

Areas of the cerebrum which receive sensory information (*sensory* areas) send impulses to **association areas**. Impulses then pass to *motor* areas, and from there to effectors. In the association area concerned with planning actions and movements, the brain integrates these sensory inputs and motor outputs to ensure that muscular movement is coordinated and appropriate. This requires the controlled action of skeletal muscles about joints.

Hint

The association area concerned with movement is in the frontal lobe of the cerebrum.

The action of skeletal muscles to produce coordinated movement can be seen in the movement of the elbow joint.

A muscle can exert force only by contracting, not by lengthening. There must be at least a pair of muscles attached across a joint, arranged to pull in opposite directions. Such muscles are called **antagonistic muscles**. When one muscle contracts, the other must be relaxed. At the elbow, contraction of the biceps and brachialis muscles causes the elbow to bend (flex). They are **flexor** muscles. Contraction of the triceps straightens the joint and extends the arm. It is an **extensor** muscle.

The arrangement and actions of the muscles of the elbow joint are shown in the diagram.

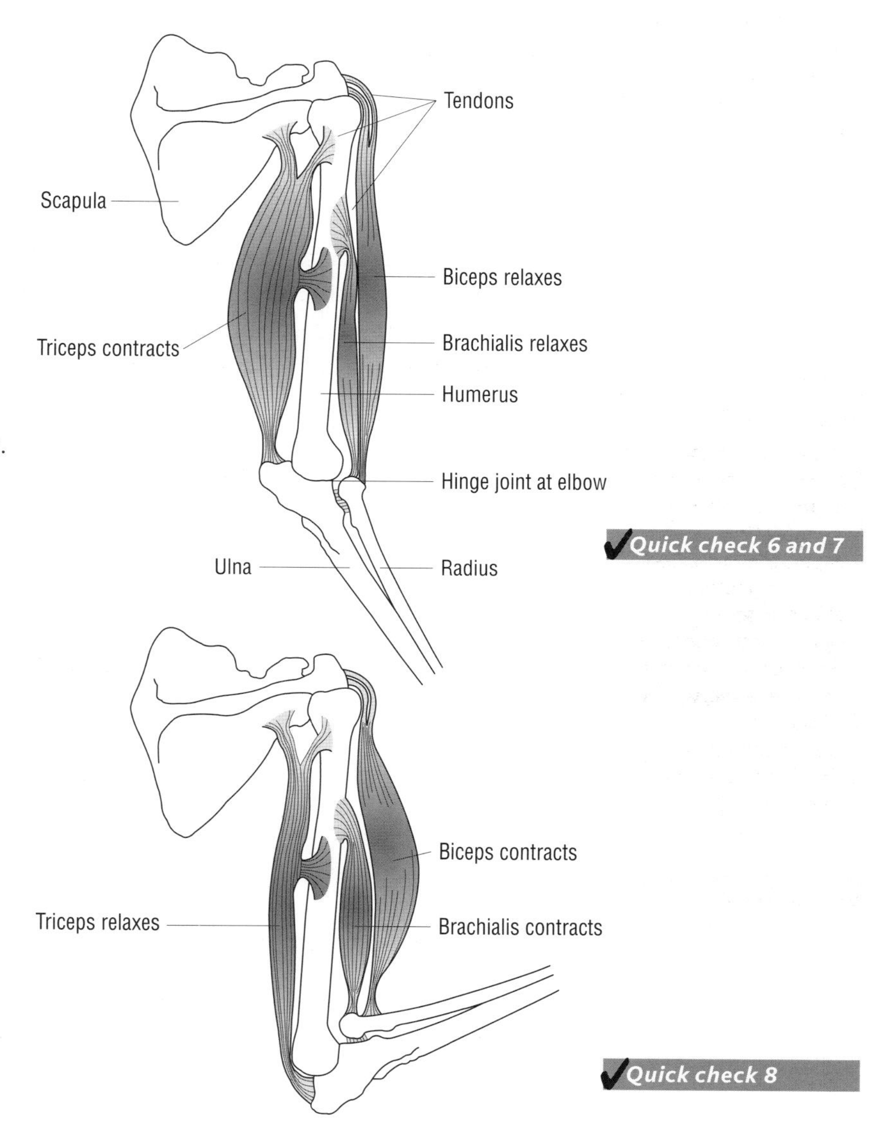

✓Quick check 6 and 7

Joints at which the two bones can move significantly with respect to one another are called *synovial* joints. The ends of the bones are covered by cartilage, and the movement is lubricated by *synovial fluid*, held within a capsule.

The joint at the elbow is a *hinge* joint, which allows movement in one plane only. Compare movement at your elbow with that at the shoulder, which is a *ball-and-socket* joint, allowing rotary movement.

✓Quick check 8

QUICK CHECK QUESTIONS

1 Distinguish between the central and peripheral nervous systems.
2 What is the somatic nervous system?
3 Explain what is meant by the autonomic nervous system.
4 Copy the longitudinal section of the brain and label its different regions.
5 State the roles of the cerebrum, cerebellum and medulla oblongata.
6 What are antagonistic muscles?
7 Distinguish between flexor and extensor muscles.
8 What muscular action is needed to flex the elbow?

Muscle contraction

Key words

- muscle fibre
- myofibril
- actin filament
- myosin filament
- sarcomere
- sarcolemma
- T-tubules
- sarcoplasmic reticulum
- creatine phosphate

Examiner tip

Skeletal muscle is also called *striated muscle* or *voluntary muscle* (see page 88).

Examiner tip

The prefix *myo-* means 'concerning muscle'.

Actin and part of myosin are fibrous proteins. Remind yourself of the structure of fibrous proteins from Unit 2 of your AS course, and of the structure of the endoplasmic reticulum from Unit 1.

✓Quick check 1, 2 and 3

Hint

Actin filaments include two other proteins: *troponin* and *tropomyosin*. In a relaxed muscle, the binding sites for myosin on the actin molecules are blocked by tropomyosin. Calcium ions bind to troponin, resulting in a shape change that lifts the tropomyosin away from the binding sites, and so act as the trigger for contraction.

✓Quick check 4

A skeletal muscle cell, or **muscle fibre**, contains many nuclei and is several centimetres long. The cytoplasm contains blocks of parallel structures (organelles) called **myofibrils**, which are bundles of thin **actin** and thick **myosin filaments**. One such block is called a **sarcomere**. The cell surface membrane, or **sarcolemma**, has inturnings called **T-tubules**, which surround the myofibrils. The endoplasmic reticulum is specialised as a **sarcoplasmic reticulum** into which calcium ions are pumped, using ATP, when the muscle is relaxed.

The diagram shows the structure of a small part of a muscle fibre, a relaxed and a contracted sarcomere. A sarcomere is a functional unit of a muscle fibre.

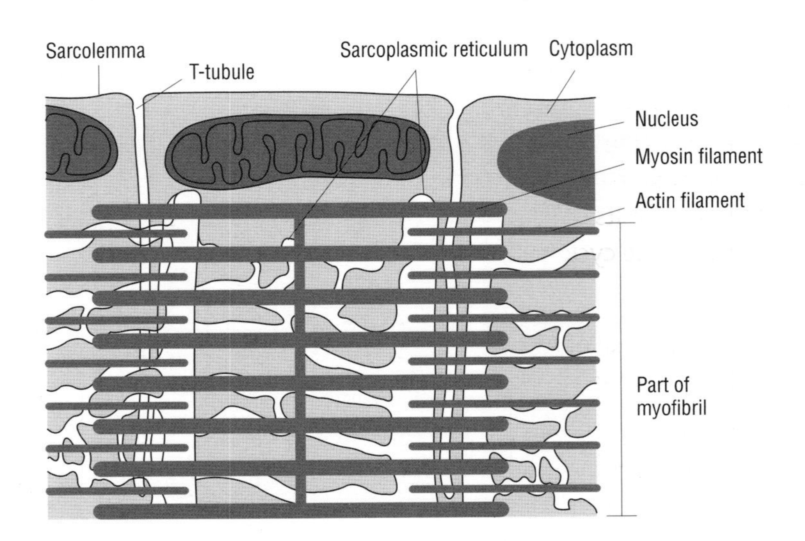

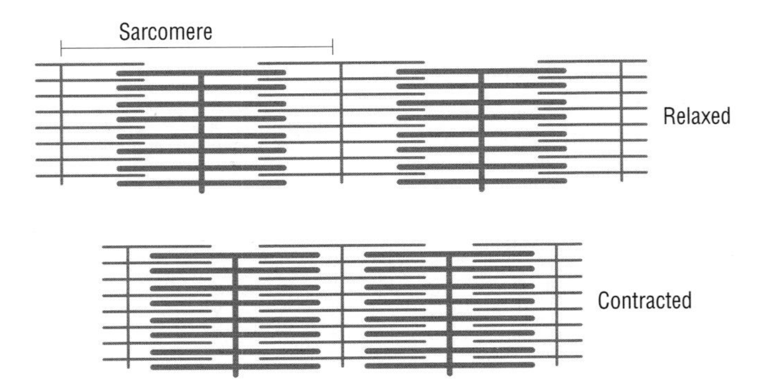

When a nerve impulse arrives at a neuromuscular junction (see page 88), the following events lead to muscle contraction.

- The cell surface membrane (sarcolemma) of the muscle fibre is depolarised, and an action potential (see page 6) spreads across it, including the membranes of the T-tubules.
- This, in turn, depolarises the membrane of the sarcoplasmic reticulum, which becomes permeable to its enclosed calcium ions.
- Calcium ions flood into the cytoplasm and bind to a protein associated with the actin filaments.
- This makes it possible for myosin to bind to actin, and for each sarcomere to shorten.

Role of ATP in muscle contraction

A sarcomere contracts by sliding the thin actin filaments over the thick myosin filaments. Myosin filaments are made up of many myosin molecules, each with a flexible 'head'. This head is an ATPase which can hydrolyse ATP to ADP and P_i (see page 27). In resting muscle, the ADP and P_i are bound to the head.

When the muscle is activated by a nerve impulse and calcium ions are released from the sarcoplasmic reticulum, the following cycle of events occurs.

- A myosin head binds to the portion of actin filament next to it. The head then tilts about 45 °, moving the attached actin filament about 10 nm in relation to the myosin, towards the centre of the sarcomere. This is the *power stroke*. The combined effect of millions of such power strokes makes the muscle contract.
- At the same time, ADP and P_i are released from the myosin.
- Then another ATP binds to the head and is hydrolysed to ADP and P_i, releasing energy that allows the myosin and actin to separate. The head tilts back to its original position, ready for the cycle to repeat.

The diagram shows the role of ATP and myosin in muscle contraction.

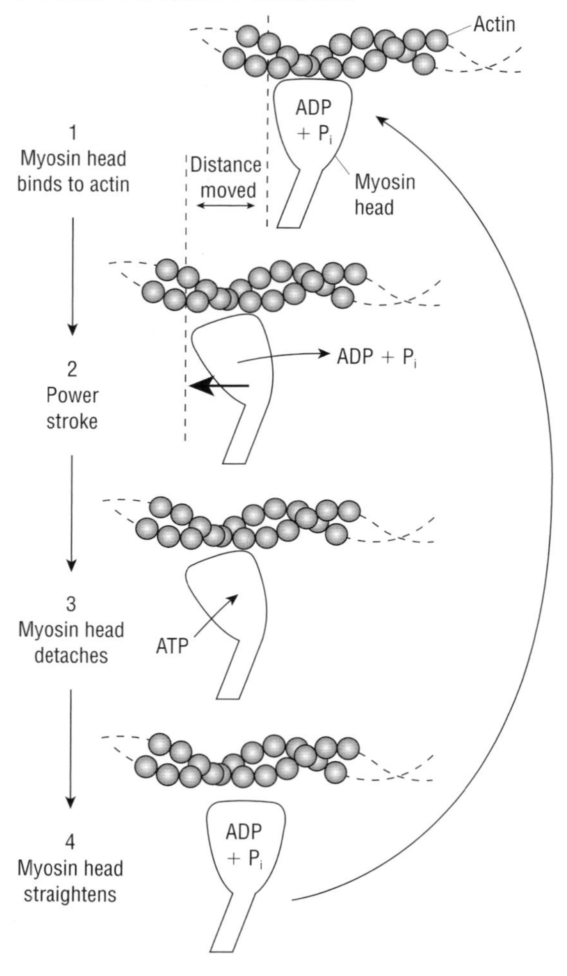

The ADP produced during muscle contraction is converted back to ATP by transferring a phosphate group from **creatine phosphate**. There is only a limited supply of creatine phosphate – once it is used up, creatine phosphate must be replenished using ATP from respiration. When a muscle is very active, the oxygen supply is insufficient to maintain aerobic respiration. The lactate pathway (see page 32) is then used to allow production of some ATP by anaerobic respiration, and the muscle cells incur an oxygen debt.

Examiner tip

Note that the hydrolysis of ATP and the power stroke do not occur at the same time. When excitation ceases, ATP is needed to actively transport calcium ions back into the sarcoplasmic reticulum.

✓Quick check 5 and 6

✓Quick check 7

Module 4

QUICK CHECK QUESTIONS

1. Distinguish between a muscle fibre, a myofibril and a filament.
2. What is the sarcoplasmic reticulum?
3. Draw and label diagrams of a relaxed and a contracted sarcomere.
4. Describe the events that follow the arrival of a nerve impulse at the sarcolemma.
5. Outline the role of ATP in muscle contraction.
6. What are the roles of a myosin head in muscle contraction?
7. What is the role of creatine phosphate in muscle contraction?

Muscle structure and function

Key words

- neuromuscular junction
- synapse
- voluntary muscle
- involuntary muscle
- cardiac muscle
- fight-or-flight response

Examiner tip

Compare the events at a neuromuscular junction with those at a synapse (see page 8).

✓Quick check 1

The diagram shows the structure of a **neuromuscular junction**. Arrival of an action potential releases vesicles of acetylcholine (ACh) from the *presynaptic membrane*. The ACh molecules diffuse across the synaptic cleft and bind to receptors on the *postsynaptic membrane* (the sarcolemma), causing depolarisation.

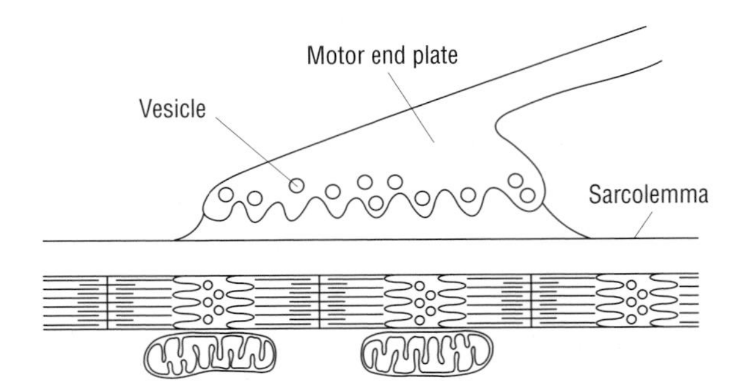

A neuromuscular junction is really a type of **synapse** (see page 8), but the two can usefully be compared.

Synapse	Neuromuscular junction
Postsynaptic membrane is cell surface membrane of a neurone.	Postsynaptic membrane is cell surface membrane (sarcolemma) of a muscle.
Neurotransmitter may be ACh, noradrenaline, glutamate or another transmitter.	Neurotransmitter is ACh for skeletal muscle.
Depolarisation of the postsynaptic membrane may be stimulatory or inhibitory.	Depolarisation of the postsynaptic membrane is stimulatory.
In both, neurotransmitter is secreted, diffuses across a cleft, binds to receptors in the postsynaptic membrane and is finally broken down.	

✓Quick check 2

Types of muscle

Three distinct types of muscle can be identified. Their structural and functional differences are summarised in the table.

Hint

The stripiness (striations) of **voluntary** and **cardiac muscle** are the result of the organisation of the myofibrils of actin and myosin (see page 86). **Involuntary muscle** contracts in the same way, but the myofibrils are not arranged in blocks.

Voluntary muscle	Involuntary muscle	Cardiac muscle
Striated (striped)	Unstriated (non-striped/ smooth)	Semi-striated
Cylindrical cells are multinucleate	Spindle-shaped cells each have a single nucleus	Cylindrical cells, each with a single nucleus, branch and connect with other cells
Found attached to bone, hence *skeletal* muscle	Found in the walls of tubular structures such as the gut, blood vessels and ducts	Found only in the heart
Controlled by the somatic nervous system	Controlled by the autonomic nervous system	Controlled by the autonomic nervous system
Contracts quickly; fatigues quickly	Contracts slowly; fatigues slowly	Contracts spontaneously without fatigue

✓Quick check 3 and 4

The fight-or-flight response

Responses to environmental stimuli in mammals are coordinated by both the nervous system and the endocrine system. The **fight-or-flight** response to an environmental stimulus is a good example of the interaction of different parts of the nervous system, together with the action of a hormone.

Imagine a dangerous change in your external environment. The following events occur.

- Sensory neurones of the somatic nervous system carry impulses from receptors to the sensory areas of the cerebrum of the brain, giving information about the danger in the environment.
- Nerve impulses pass to association areas in the cerebrum.
- Nerve impulses in sympathetic nerves of the autonomic nervous system, from the brain to the sinoatrial node (SAN) of the heart, increase the pulse rate and the stroke volume of the heart.
- Impulses in sympathetic nerves from the brain to the adrenal glands cause secretion of adrenaline from the adrenal medulla (see page 11).

Adrenaline secreted into the blood stream then has a number of effects, including:

- stimulation of the heart, with the same effects as stimulation by sympathetic nerves
- increase in blood pressure, by constriction of blood vessels to the skin and gut
- increase in air flow to the lungs
- increased breakdown of glycogen in the liver
- decreased sensory threshold and increased mental awareness.

The actions of adrenaline provide an increased flow of oxygenated blood carrying glucose. With the body prepared in this way for the needs of muscles, which may work hard for the organism to escape from or cope with a source of danger, a decision is made about how to respond.

- Nerve impulses from the association area in the frontal lobe of the cerebrum (the prefrontal association complex), concerned with planning actions and movements, pass to motor areas.
- From there, motor neurones of the somatic nervous system carry impulses to muscles, to produce the chosen action.

Examiner tip

Review coordination of the heartbeat from Unit 1 of your AS course.

Examiner tip

Remind yourself of the hydrolysis of glycogen to glucose from Unit 2 of your AS course.

Note that adrenaline has the same effect as glucagon on the liver (see page 12).

Quick check 5 and 6

Quick check 7

Module 4

QUICK CHECK QUESTIONS

1. List the events triggered by a nerve impulse arriving at a neuromuscular junction.
2. State two similarities and two differences between a synapse and a neuromuscular junction.
3. State one structural and one functional difference between voluntary muscle and cardiac muscle.
4. Where is involuntary muscle found in the body?
5. Describe the way in which nerve impulses in the sympathetic nervous system prepare the body for fight-or-flight.
6. How does secretion of adrenaline prepare the body for muscular activity?
7. What is the role of the cerebrum in the fight-or-flight response?

Animal behaviour

Animals behave in ways that enhance their survival and reproductive capacity. Behaviour patterns can be simple or complex, and can range from genetically programmed behaviour to learned behaviour that is significantly influenced by the environment.

Key words

- innate
- learned
- escape reflex
- kinesis
- taxis
- habituation
- imprinting
- classical conditioning
- conditioned reflex
- unconditioned stimulus
- conditioned stimulus
- operant conditioning
- latent learning
- insight learning

Animal behaviour may be **innate** or **learned**.

- Innate (or instinctive) behaviour is a pattern of genetically determined behaviour that does not require learning or practice.
- Learned behaviour involves an adaptive change in response – behaviour based on experience.

Innate behaviour

Examples of innate behaviour include the following.

- **Escape reflex** – in which a particular stimulus brings about an automatic response. For example, earthworms contract the longitudinal muscles of the body when they are touched at the head end, and so withdraw into their burrow to escape predation.
- **Kinesis** – a movement in response to an external stimulus, in which the rate of movement is related to the intensity, but not the direction, of a stimulus. Woodlice move rapidly and turn frequently in dry conditions. When damper conditions are found *by chance*, movement slows down or stops, keeping the organism within optimal conditions.
- **Taxis** – a directional movement in response to an external stimulus. Woodlice move *away* from light: a *negative* phototaxis. They will be less visible to predators in darker conditions, and less liable to desiccation.

✓Quick check 1

Module 4

Hint

The plural of *taxis* is *taxes*; the plural of *kinesis* is *kineses*.

The advantages of innate behaviour are:

- it does not need to be learned
- it has immediate survival value for a young, inexperienced animal in a dangerous situation
- it is appropriate for invertebrates with short life cycles that do not have time to learn
- it requires few neurones
- it is likely to be appropriate for the animal's habitat, as the alleles controlling it will have been subject to natural selection.

✓Quick check 2 and 3

Learned behaviour

Learned behaviour includes the following types of behaviour.

- **Habituation** – the simplest form of learning, in which an animal stops responding to a repeated stimulus. In this way, an animal learns to ignore a stimulus that is no longer novel and is neither harmful nor beneficial. The earthworm's escape response fades, allowing it to emerge from a partially blocked burrow. Calcium ion channels in presynaptic membranes are inactivated so that less and less neurotransmitter is secreted.
- **Imprinting** – a form of learning to recognise a parent or other complex stimulus, which is often limited to a sensitive period in an animal's development. It ensures that the infant animal is in a position to learn the skills possessed by the parent. There is no innate recognition of the parent species: recently hatched orphan greylag goslings learn to recognise and follow the patterned boots of their surrogate human 'parent'.

✓Quick check 4

- **Classical conditioning** – a form of adaptive learning in which an innate response is modified. The animal learns to respond to a stimulus that is different from the usual stimulus, as seen in Pavlov's dogs salivating in response to a bell, rather than to the presence of food, in a **conditioned reflex**. For his classic series of experiments, Pavlov inserted a tube into a dog's cheek so that the saliva produced could flow out and its volume could be measured. Salivation in response to the taste or smell of food is an innate response, and the stimulus producing it is an **unconditioned stimulus**. In controlled conditions, the dog was given a standard quantity of food and the volume of saliva produced was measured. No saliva was produced when a bell was rung and no food given. But after several days on which a bell was rung before the food sample was given, it was found that the dog salivated at the sound of the bell alone. The new stimulus is called a **conditioned stimulus**. The dog associated the sound of the bell (the conditioned stimulus) with the arrival of food (the unconditioned stimulus) so that either produces a response.
- **Operant conditioning** – a form of adaptive learning in which an animal learns to carry out a particular action in order to receive a reward or to avoid an unpleasant experience. An example is a rat in a Skinner box pressing a lever to be fed, or to avoid an electric shock. When an animal is first put into such a box, it explores. It moves around looking for a way out. It may accidentally press the lever and be rewarded. After a few such experiences, the animal presses the lever in order to receive the reward. It has learned to associate the reward with that action.
- **Latent (exploratory) learning** – behaviour that is not directed towards a particular outcome. Animals explore new surroundings and learn information that has no apparent value at the time, but may be useful at some time. Mice may more easily escape from a predator if they know the surroundings of their hole.
- **Insight learning** – a form of learning in which an animal integrates memories of *two or more* earlier actions (e.g. from latent learning) to produce a new response to gain a reward. Chimpanzees provide good examples of such behaviour. A group of chimpanzees had played with a number of cardboard boxes, sometimes putting one on top of another and climbing on them. They also had experience of standing on one box to reach bananas that were otherwise out of reach. When they were then presented with bananas hung well out of reach, they were able to use their past experiences and make a pile of boxes to reach the fruit.

Hint

A Skinner box is an animal cage with a lever which, when pressed, releases a reward of food or switches off an unpleasant stimulus.

Hint

Both classical conditioning and operant conditioning are forms of associative learning.

Examiner tip

Insight learning is more complex than the associative learning seen in classical or operant conditioning, but is it evidence of the animal thinking about how to solve a problem?

✓*Quick check 5, 6 and 7*

QUICK CHECK QUESTIONS

1. Distinguish between innate and learned behaviour.
2. Name three types of innate behaviour.
3. State three advantages to an animal of innate behaviour
4. Explain what is meant by habituation and imprinting.
5. Distinguish between classical and operant conditioning.
6. What is meant by latent learning?
7. Describe an example of insight learning.

Social behaviour in primates

Key words
- social behaviour
- dopamine
- DRD4

Animals rarely live alone, and many show a variety of behavioural activities associated with members of a species living together, temporarily or permanently, in a group. Such behaviour is **social behaviour**.

✓*Quick check 1*

Social behaviour in humans is developed to a greater extent than in any other species, but all monkeys and apes are sociable. Most live in groups of related females and their offspring, accompanied by one or more males. Individuals in social groups often assume specialised roles, as seen in the social insects such as bees or ants. This increases the efficiency of the group. However, primate societies are flexible, and many roles are interchangeable. How did higher primates become sociable? Primates tend to mature slowly and late, and to live for a long time. In these circumstances, the basic needs of food, safety and reproduction are provided most effectively by group activities. Learned behaviour plays an important role in cooperation between members of the group. A communication system is essential for cooperation, and the most advanced primate societies have the least stereotyped and most varied visual and vocal signals.
The advantages of social behaviour involving cooperation are clearly seen among primates. An individual chimpanzee is unlikely to be able to scare away a predator, but a group of chimpanzees can learn to cooperate in throwing sticks and stones to see off a leopard.

Male chimpanzees in the Ivory Coast's Taï forest show social cooperation in hunting prey. Adult males move through the forest in search of red colobus monkeys feeding in the tree canopy. Once a group of monkeys is found, the chimpanzees spread out and surround a victim, cutting off its possible escape routes. Then some of the males move in to capture the monkey. A successful hunt depends on each chimpanzee responding effectively to the movements of each of its companions, as well as to those of the monkey trying to escape. This involves a high level of social cooperation and learning. Not all groups of chimpanzees hunt in the same way – different techniques are learned by different groups.

✓*Quick check 2 and 3*

Hint

This type of natural selection is called *kin selection*.

Such social behaviour is a balance of individual self-interests. The behaviour patterns seen in animals living in social groups are likely to contribute to an individual passing on his or her alleles to the next generation. An animal that cooperates with a relative is increasing the chance that the alleles they have in common will survive. The greater the genetic overlap between individuals, the greater the incentive to cooperate rather than to compete.

✓*Quick check 4*

Dopamine receptors and behaviour

Genetic differences between individuals can be linked to differences in behaviour. Different alleles result in variation in the physical structure of an individual, which in turn affects the individual's behaviour in particular circumstances. This can be illustrated by the variations in behaviour linked to the possession of different receptors in the brain for the neurotransmitter, **dopamine**. Different receptors have different abilities to bind dopamine, affecting brain activity.

Dopamine, which enhances general arousal, is made from the amino acid tyrosine. It is associated particularly with voluntary movements. People with low concentrations of dopamine have difficulty in initiating movement, as in Parkinson's disease.

The gene coding for one of the five receptors for dopamine, **DRD4**, is one of the most variable human genes known. A 48 base pair sequence may be repeated up to 11 times. Each change to the primary structure of the protein results in a change in the three-dimensional shape of the receptor, which alters its ability to bind dopamine. As a result of this, the different alleles appear to be responsible for individual differences in susceptibility to neuropsychiatric disease and in responsiveness to medication. For example:

- the allele with seven repeats, *DRD4 7R*, is associated with higher than average test scores for novelty-seeking
- the *DRD4 7R* allele is also associated with attention deficit–hyperactivity disorder (ADHD)
- the allele with five repeats, *DRD4 5R*, is associated with alcoholism and other drug abuse
- some studies have found links between DRD4 alleles and complex disorders such as schizophrenia.

The link between a particular *DRD4* genotype and a mental state such as schizophrenia is a statistical association. Many genetic and environmental factors apart from the different *DRD4* alleles are involved. Understanding the role of the DRD4 protein receptors and other neuroreceptors is helping the development of therapeutic drugs.

Examiner tip

Review protein structure from Unit 2 of your AS course.

Hint

Note that repeating a 48 base pair sequence of DNA five times adds 80 amino acids to the primary structure of the protein. Repeating it seven times adds 112 amino acids (see page 43).

Quick check 5, 6 and 7

QUICK CHECK QUESTIONS

1. What is meant by social behaviour?
2. Suggest advantages of primates living in social groups.
3. Describe two examples of cooperation in chimpanzees.
4. Explain the selective advantage of cooperation in a related group of animals.
5. Describe two different alleles of the gene coding for the dopamine receptor DRD4.
6. Explain why different alleles of the gene for DRD4 result in different abilities of the receptor to bind dopamine.
7. State one example of a genetic difference between individuals that can be linked to differences in behaviour.

End-of-unit questions

1 For protein synthesis to be possible, each of the different types of amino acid involved must first combine with the appropriate tRNA molecules, as shown in the two equations:

tRNA + amino acid + ATP + enzyme → enzyme–substrate complex
enzyme–substrate complex → enzyme + ADP + phosphate + amino acid linked to tRNA

(a) (i) Suggest why ATP is needed to enable an amino acid molecule to combine with a tRNA molecule. (1)

(ii) Explain why amino acid molecules must combine with tRNA molecules for a specific protein to be produced. (4)

Liver cells were placed in a solution that contained radioactive amino acids 'labelled' with ^{14}C, the radioactive isotope of carbon. Organelles from these cells were then extracted, and the radioactivity of each type of organelle was measured.

(b) (i) Which of the following organelles would be the first to become radioactive?
nucleus, Golgi apparatus, smooth endoplasmic reticulum, rough endoplasmic reticulum, mitochondrion (1)

(ii) Explain your choice. (2)

2 Mutations are random changes to nucleotide sequences within a cell. The sequence may be altered in the ways shown below.

Type of random change	Original sequence	New sequence following change
Deletion of one or more nucleotides	AAAG**T**GATCCCC	AAAGGATCCCC
Addition of one or more nucleotides	AAAGTGATCCCC	AAAGTGATCC**G**CC
Substitution of one or more different nucleotides for ones in the previous sequence	AAAG**T**GATCCCC	AAAG**A**GATCCCC
Inversion – the same nucleotides but in a new sequence	AAAGT**GAT**CCCC	AAAGT**TAG**CCCC

The change in nucleotide sequence *may* result in the synthesis of a protein with a different amino acid sequence from the protein that the cell had previously produced.

(a) Explain why a random change would have a much greater effect on an organism if it occurred in DNA, rather than in RNA. (2)

(b) (i) Suggest why substitution mutations often make no difference to the protein molecule coded by the DNA. (3)

(ii) Explain why the inversion of GAT to produce TAG would be likely to result in a different protein. (1)

(iii) Suggest why the deletion or addition of three nucleotides from a gene usually results in a much smaller change to the protein coded than does the deletion or addition of a single nucleotide. (3)

3 Groundsel is a common weed. Groundsel plants vary in the structure of their flowers. Three distinctive phenotypes can be recognised and are shown in the figure.

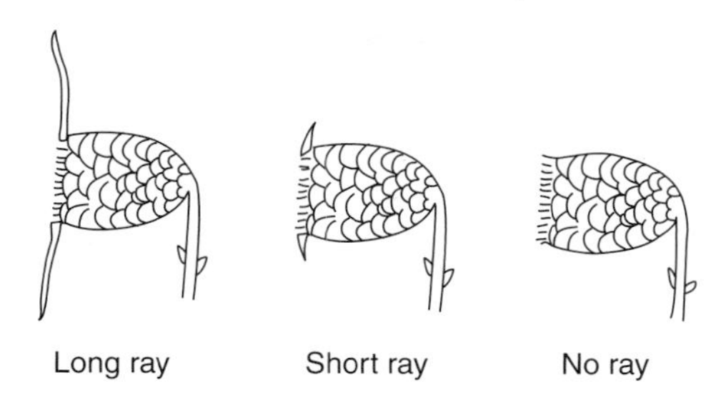

Magnification ×1.5

All flowers on any one plant are always of the same type. Seedlings of groundsel were collected from a population known to contain all three types. The seedlings were grown to maturity in individual pots under conditions where cross-pollination would be unlikely. The seeds produced by each of the resulting plants were collected and sown separately. When the offspring had grown to maturity, each was examined and its flower type noted. The results are shown in the table below.

Phenotype of parent plant (ray type)	Number of offspring of each parent with phenotype (ray type)		
	Long	Short	None
1. Long	103	0	0
2. Long	89	0	0
3. Short	38	87	41
4. Long	23	0	0
5. None	0	0	101
6. Long	121	1	0
7. None	0	0	76
8. Short	53	111	61
9. Long	153	0	0
10. Short	23	42	19

With reference to the table, answer the following questions.

(a) Describe the trends shown by the data. (3)

(b) Explain the results using suitable genetic symbols and diagrams. (5)

(c) Predict the offspring that would result from a cross between a long-rayed plant and one with no rays. (1)

(d) State *two* hypotheses to explain the single short-rayed offspring produced by parent number 6; assume this plant was correctly identified as short-rayed. (2)

4 Mature, ripe pea seeds vary in shape. Some commercial varieties have smooth, almost round seeds. Other varieties have more irregular, wrinkled seeds. This variation is inherited and the pattern of inheritance shows that one gene is involved. A dominant allele controls the development of round seeds, and a recessive allele controls the development of wrinkled seeds.

Examiner tip

Question 4 is synoptic.

Examiner tip

You don't need to know the genotype of the other parent. (Why not?)

A plant with wrinkled seeds was cross-pollinated, and one of its offspring had round seeds.

(a) Predict the ratio of offspring that would be produced if this round-seeded plant was cross-pollinated by a plant with wrinkled seeds. (1)

A round seed and a wrinkled seed were crushed in water, and a random sample of starch grains from each seed was observed using a microscope. The grains from the round seed were usually oval in shape, while most of those from the wrinkled seed were triangular. The grains were measured at their widest point. The data are recorded in the table below.

Maximum diameter of starch grain/μm									
From round seed	32	29	18	23	22	31	28	26	21
From wrinkled seed	6	7	5	7	7	5	8	5	6

(b) With reference to the table, calculate the mean diameter for each type of starch grain. (2)

(c) Suggest how random grains might be selected for measurement using a microscope. (1)

(d) State how the grains could be identified as containing starch. (1)

(e) Explain how it is possible for the size of starch grains in a seed to vary, even though all cells in the seed have the same genotype. (2)

The figure shows glucose 1-phosphate.

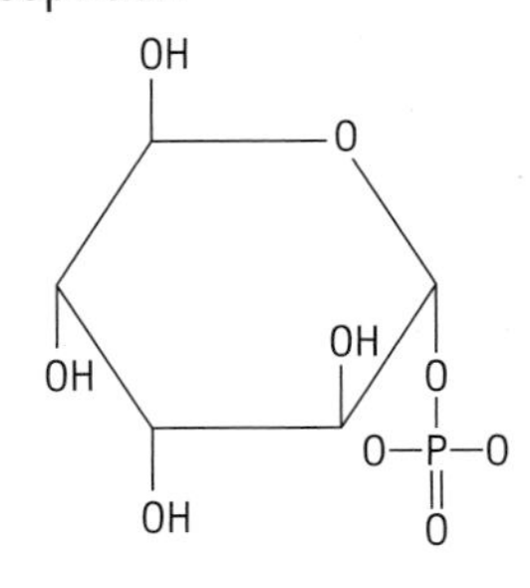

Glucose-1-phosphate

(f) With reference to the figure, state whether the glucose in this substance is α or β glucose. (1)

Examiner tip

Think about the properties of enzymes and apply your knowledge of enzyme characteristics to this example.

A starch-free solution, rich in soluble proteins, can be extracted from pea seeds. A few drops of this extract will produce starch if added to a solution of glucose-1-phosphate. No starch is produced by the extract if it is added to glucose or to maltose solutions.

(g) Suggest how it could be demonstrated that pea seed extract contains an enzyme that controls the formation of starch from glucose-1-phosphate. (1)

(h) Assuming that starch is synthesised from glucose-1-phosphate within developing pea cells, suggest what substance reacts with glucose to produce the glucose-1-phosphate in the cells. (1)

Extracts made from round pea seeds are always more active in the synthesis of starch than extracts made from wrinkled peas.

(i) Describe *two* experimental variables that would need to be controlled when comparing the enzyme activity of extracts from round and wrinkled peas. (2)

(j) Explain the relationship between the enzyme activity of pea seeds and the genotype of the seeds. (2)

5 Elm trees reproduce by growing shoots from an existing root system. Such shoots are called suckers. A single tree can produce a chain of offspring of various ages in this way. Oak trees all grow from seeds called acorns. They do not reproduce by means of suckers.

(a) Explain why you would expect the genetic diversity of a population of oaks to be much greater than the genetic diversity of a population of elms. (4)

The oil palm (*Elaeis guineensis*) is grown in tropical areas and is of increasing economic importance. At the centre of the very large leaves that crown the oil palm tree is a group of cells from which new leaves develop. This apical meristem can be cut out of a tree and divided with a sterile scalpel into many small pieces. Each of these pieces will grow into a plant if it is placed on suitable agar.

(b) Discuss the advantages and disadvantages of propagating palm oil trees in this way. (4)

(c) Compare the methods used to produce artificial clones of mammals and plants. (7)

Examiner tip

For 5(b), make clear which are advantages and which disadvantages.

Examiner tip

Make sure you *compare* cloning in mammals and plants for 5(c). Don't describe first one and then the other. *Compare* means look for similarities and differences.

6 Trimethylamine, $N(CH_3)_3$, is produced during decay, and is one of the substances that make rotting fish and sewage stink! Some species of bacteria produce an enzyme that breaks down trimethylamine. This enzyme was investigated to see if it could be used to decontaminate water as part of sewage treatment. Bacteria producing the enzyme were cultured in a fermenter. This had electronic sensors that monitored the solution within.

(a) Explain why a pH-monitoring sensor is needed, and outline how it operates. (4)

Examiner tip

There are two parts to question 6(a): *outline* means select the main points.

Bacteria were harvested from the fermenter and lysed to release the enzyme. The enzyme was immobilised in alginate beads. Water discharged during sewage treatment was passed through tubes packed with the beads, so that trimethylamine could be broken down.

(b) (i) Explain why the enzyme solution could not be simply added to the sewage effluent. (2)

(ii) Suggest advantages and disadvantages of using larger or smaller beads of immobilised enzyme. (3)

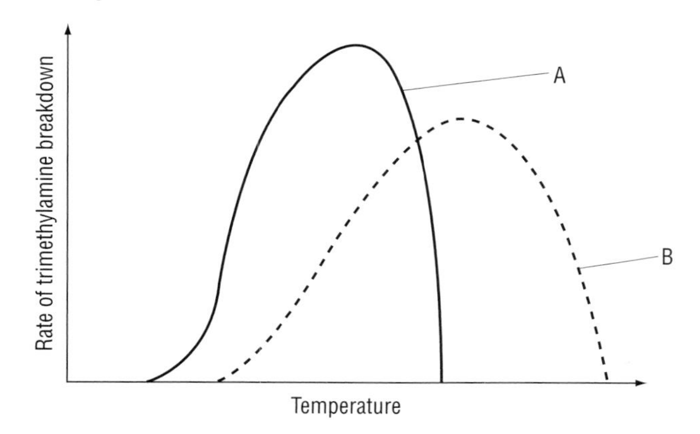

The figure shows how the rate of trimethylamine breakdown varies with temperature for a dissolved enzyme (A), and using the same concentration of the enzyme in an immobilised form (B).

(c) (i) Describe the differences between curves A and B. (3)

(ii) Explain these differences. (3)

Examiner tip

This question is partly synoptic.

7 The diagram shows a plasmid that includes antibiotic resistance genes, labelled gene 1 and gene 2.

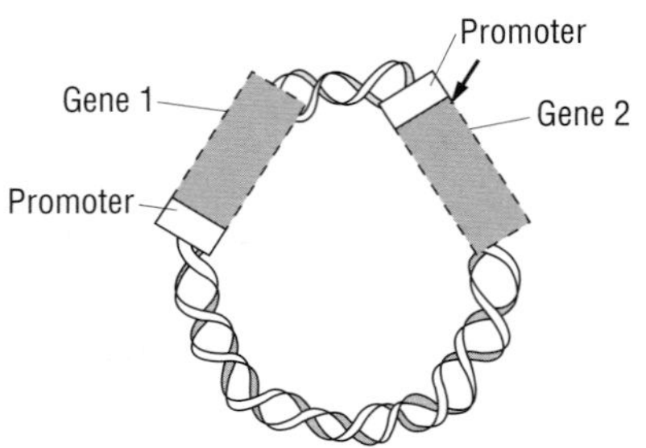

Gene 1 gives resistance to antibiotic A; gene 2 gives resistance to antibiotic B. When using this plasmid to introduce a new gene into bacteria, a strain of bacteria is chosen that are harmless, easy to grow in a fermenter, and free of the plasmid. The plasmid is cut, using a ...., at a site indicated by the arrow, just after the promoter region of gene 2. The new gene inserts itself into the plasmid, because the cut plasmid and the new gene have complementary .... The joins between the new gene and plasmid are made secure using the enzyme DNA .... The recombinant plasmid and the bacteria are mixed and the solution is spread over agar jelly containing antibiotic A.

Examiner tip

You will need to remember what a promoter is. Think of the *lac* operon example.

With reference to the figure and the passage, answer the following questions.

(a) Write down an appropriate word or words to fill each gap in the passage. (3)

(b) All the cells in a bacterial colony that is able to grow on the agar must have been produced by division of a cell that had absorbed a plasmid. Explain why. (2)

(c) Some of the colonies that develop on the agar are sensitive to antibiotic B. Explain why. (1)

(d) Suggest why the new gene is inserted into the plasmid very near the promoter of gene 2. (2)

8 *Artemisia tridentata* is a desert shrub that is known for its ability to tolerate very high temperatures. There is great concern internationally that global warming will damage rice and maize crops, which are much less tolerant of high temperatures than *A. tridentata*. If the cell membranes of *Artemisia* have a protein that makes them more stable at high temperature, the gene coding for this protein might be introduced into the genome of rice or maize, making the resulting plants tolerant of high temperature.

(a) Explain why *Artemisia* genes cannot be introduced into the rice genome by conventional means. (1)

(b) Describe how *Agrobacterium tumefaciens* might be used to introduce the *Artemisia* gene into rice plants. (5)

(c) Give two reasons why some people regard the development of genetically engineered crops as unacceptable. (2)

Another plant from the genus *Artemisia*, *A. annua*, produces a secondary metabolite called artemisinin, which is one of the most important drugs used in the treatment of chloroquin-resistant malaria.

(d) Explain the terms:

(i) secondary metabolite (2)

(ii) genus. (1)

9 Production of zinc, lead and cadmium takes place on a large scale at Avonmouth, near Bristol. Cadmium is a toxic heavy metal. It is estimated that about 3.5 tonnes per year of cadmium once escaped into the atmosphere from the Avonmouth metal smelter. Most of this was deposited onto soil and vegetation within a few kilometres of the smelter.

(a) State an effect of heavy metal ions, such as lead and cadmium, on living cells. (1)

Examiner tip

This question is synoptic.

The effect of the Avonmouth pollution was studied by comparing two oak woodlands. Hallen Wood is 3 km from the smelter and Wetmoor Wood is 23 km from the smelter.

The table shows the population density of animals inhabiting the litter and surface layers of soil and the dry mass of litter (dead plant remains) in these woodlands. The types of animal listed are detritivores, feeding mainly on dead and decaying leaves and twigs.

	Mean number of individuals per m^2		Ratio Wetmoor : Hallen (to nearest whole number)
Animals	Hallen Wood	Wetmoor Wood	
Earthworms	17	75	4 : 1
Millipedes	8	79	10 : 1
Woodlice	56	171	
Mites	129 000	194 000	1 : 1
Springtails	20 800	8688	
	Mean dry mass of litter/kg m^{-2}		
Mass of litter	14.28	1.35	1 : 11

(b) Complete the table by inserting the appropriate ratios for woodlice and for springtails. (2)

(c) With reference to the table, answer the following questions.

(i) Explain the difference in the mass of litter that has accumulated in Hallen Wood compared with Wetmoor Wood. (1)

(ii) Suggest why the ratio for springtails is different from the ratios for the other types of animal. (2)

The following table shows the concentration of cadmium in the bodies of several organisms collected in Hallen Wood, and in the litter from the wood.

	Concentration of cadmium/mg kg^{-1} dry mass
Litter	23.5
Grass (*Holcus lanatus*)	5.0
Grass (*Milium effusum*)	3.8
Bluebell	14.2
Slug (*Arion hortensis*)	56.5
Slug (*Arion alter*)	119.3
Woodlouse (*Oniscus asellus*)	125.7
Woodlouse (*Porcellio scaber*)	41.1
Earthworm	56.6

(d) With reference to the second table, answer the following questions.

(i) Calculate the mean concentrations of cadmium in the living plants and in the animals. (2)

(ii) Explain why the concentration of cadmium is different in the plants and animals. (2)

(iii) Suggest why pairs of similar species of animal, e.g. the slugs *Arion hortensis* and *A. alter*, have different cadmium concentrations. (2)

(e) Describe briefly how you would measure the abundance of bluebells, or another small plant species, in a community such as woodland. (2)

When seedlings or cuttings are grown in solutions of nutrient salts to which small quantities of cadmium have been added, root growth is poor and they fail to thrive. When this technique was used to compare the populations of the grass *Holcus lanatus* found in Hallen Wood and Wetmoor Wood, the plants from both woods grew equally well in the absence of cadmium. Those from Hallen Wood grew more strongly and with healthier roots than those from Wetmoor Wood when cadmium was present.

(f) Explain how the population in Hallen Wood has become relatively tolerant of cadmium. (7)

Examiner tip

In this question, 1 mark is available for the quality of written communication.

10 Oaks are the most important trees in natural woodlands in Britain. Two ancient woodlands have been studied by biologists from Oxford University for more than 60 years. More than 25 000 species of animal and plant have been recorded from these woodlands.

In the woodlands, different parts of oak trees form the basis for very many food chains. One of these food chains is:

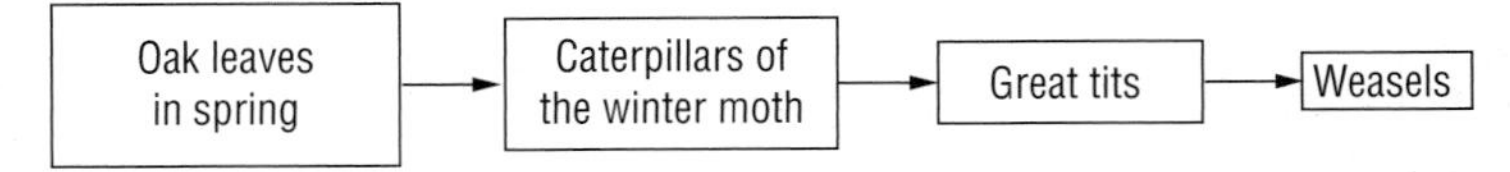

The winter moth lays its eggs on the oak buds in winter. Caterpillars hatch from these eggs in spring, after the leaves have begun to develop. The tits depend on the caterpillars as food for their young. Fully grown caterpillars drop off the trees and pupate in the leaf litter. Great tits do not forage in the leaf litter at the base of trees.

(a) (i) Explain why the winter moth eggs must hatch after the leaves have started to grow. (1)

(ii) Suggest two abiotic factors that change in spring, and that might stimulate hatching of the caterpillars. (2)

(b) Explain why very little of the energy stored in the tissues of the caterpillars ends up stored in the tissues of the tits. (4)

The graphs show some of the trends shown by data from the years of study of these tit populations.

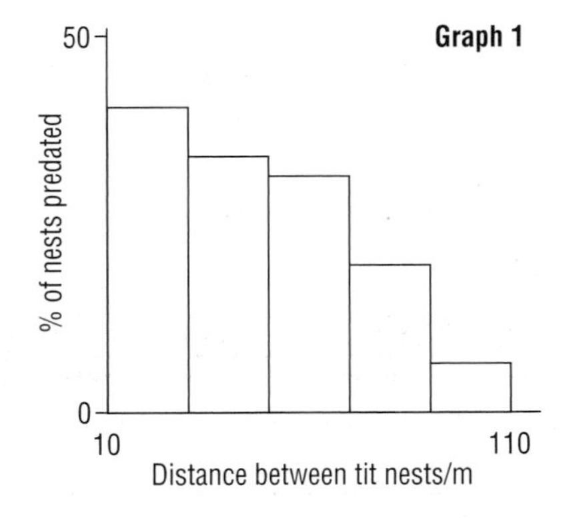

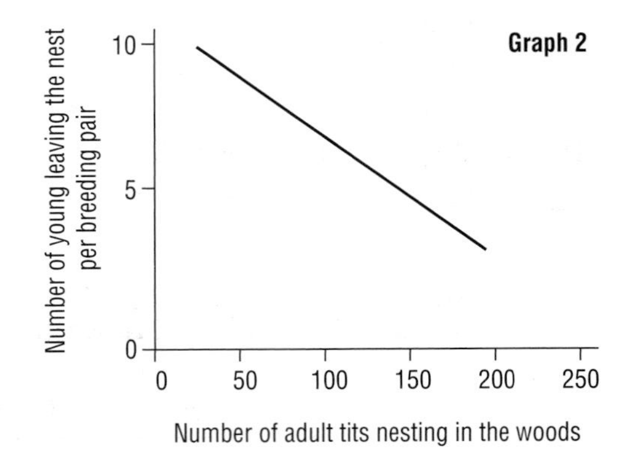

Graph 1 shows the relationship between the distance separating tit nests and the chance of the young in the nest being eaten by a weasel. Graph 2 shows the relationship between the total number of tits in the woods and the mean number of young tits successfully reared by each pair.

(c) (i) Describe and explain the trend shown by graph 1. (3)

(ii) With the help of graph 2, explain the term *carrying capacity*. Estimate the carrying capacity for tits in the woodlands studied. (2)

Large numbers of nest boxes were provided in the woods, so that researchers could examine the nests easily.

(d) Suggest how this is also a conservation measure. (2)

11 A mutation in tomatoes results in short, bushy plants. Varieties homozygous for the mutant allele are sold to grow in pots or window boxes. The short, bushy phenotype is referred to as dwarf.

Healthy, rapidly growing, young dwarf plants were treated with two different plant hormones, A and B. The way in which the hormone was applied to the plants, and the effect on the growth and development of the plants, are indicated in the table.

Treatment		Effect on development	
		Mean plant height/cm	Mean number of branches per plant
1	No treatment (control group)	36	15
2	All terminal buds removed, no hormone applied	22	28
3	All terminal buds removed and replaced by a paste containing hormone A	24	26
4	All terminal buds removed and replaced by a paste containing hormone B	32	14
5	Terminal buds *not* removed, leaves sprayed weekly with a solution containing hormone A	128	16
6	Terminal buds *not* removed, leaves sprayed weekly with a solution containing hormone B	25	13
		Leaves developed with curled, distorted shape; roots started to grow from some stems	

Examiner tip

Understanding what a control experiment shows is often essential when you evaluate and draw conclusions.

(a) Suggest why dwarf genotypes are recommended for window boxes and pots. (1)

(b) Explain why the control group of plants was included in the investigation. (2)

(c) State the effect of removing terminal buds on growth and development of the plants. (2)

(d) Compare the effect of the two hormones, A and B, when they were applied in a paste that replaced the terminal buds. (4)

(e) Explain why the greater height of the plants treated with a spray containing hormone A would *not* be passed on to their offspring. (1)

(f) Identify hormones A and B, explaining your answer. (4)

(g) Make a table like the one below to show two commercial uses for plant hormones. An example has been completed for you. (2)

Hormone	Commercial use
Auxin	Stimulates the development of roots in cuttings during artificial asexual reproduction

12 Dopamine is a substance that acts as a neurotransmitter in important brain processes, including those involved in control of movement and in mental states such as curiosity and risk-taking. Diseases such as schizophrenia and Parkinson's disease are associated with abnormal concentrations of dopamine in different parts of the brain. One of the receptors for dopamine is a protein called DRD4. There are several alleles of the gene that codes for DRD4, so that different people may inherit slightly different DRD4 structures.

(a) State the meaning of the term *neurotransmitter*. (3)

(b) Explain why a finding that 'the brains of people with schizophrenia have abnormal dopamine concentrations' does not prove that an abnormal concentration of dopamine is a cause of schizophrenia. (2)

(c) Explain why it is likely that genetic variation in DRD4 is a factor involved in variation in behaviour and personality. (2)

Amphetamines are a group of chemical compounds that have structural features similar to dopamine. Taking amphetamines produces feelings of excitement, but often results in mental illness.

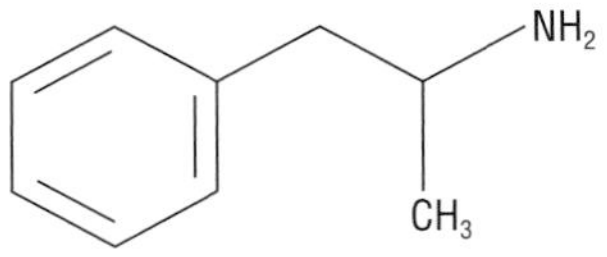

NH_2 CH_3 HO HO NH_2

Amphetamine 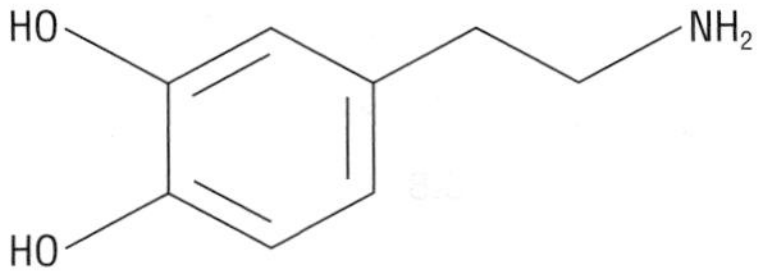 Dopamine

(d) Use the molecular diagrams to suggest why amphetamines have these effects. (2)

13 A male rat is able to mate successfully with a receptive female when he has been hand-reared and has never before seen or come into contact with any other rat. Hand-reared male monkeys are rarely able to mate with receptive females. Monkeys require prolonged interaction with others in a social group to develop fully functional sexual behaviour.

(a) (i) Suggest an explanation for this difference between male rats and male monkeys. (2)

(ii) Apart from enabling the development of sexual behaviour, describe *one* other advantage to monkeys of living in a social group. (1)

The scientist Konrad Lorenz hand-reared a male house sparrow. When it became adult, the sparrow attempted to mate with Lorenz's hand and displayed to the hand, just as wild male sparrows display when courting a female sparrow.

(b) State the type of behaviour that was shown by this sparrow, and explain the advantage of this type of behaviour pattern to wild sparrows. (2)

Honeybees are social insects. When they find a glass dish of sucrose solution, they visit it repeatedly until it has all been carried back to their colony. The scientist Karl von Frisch fed bees in this way, always placing the dish on a blue tile. When the sucrose was all consumed, the bees searched empty glass dishes, but only if these were placed on a blue background.

(c) (i) Explain why the behaviour of von Frisch's bees is an example of classical conditioning rather than operant conditioning. (2)

(ii) Karl von Frisch was able to show that bees can see blue, but not red. Suggest, in outline, how he might have carried out his investigation. (4)

14 The figure shows how the frequency of three diploid genotypes, **AA**, **Aa** and **aa**, change with changes in the frequency of the recessive allele, **a**.

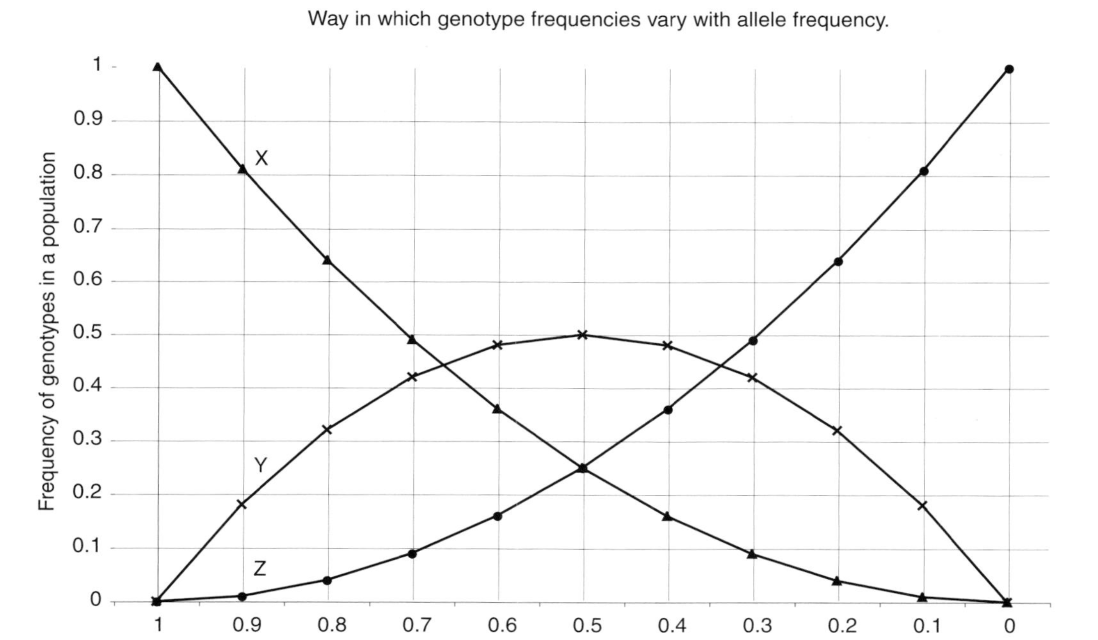

(a) (i) Which curve, **X**, **Y** or **Z**, represents the frequency of genotype **Aa**? (1)

(ii) How would the horizontal axis scale change if it were to show the frequency of the dominant allele, **A**? (1)

Most wild primroses (*Primula vulgaris*) have yellow flowers. Some populations of primroses show variation in flower colour, a minority of the plants having pink flowers. Genetic analysis shows that the pink-flowered phenotype is the result of a dominant allele (**A**). The yellow flower is the result of a recessive allele (**a**).

All the primroses flowering in a small wood were counted. There were 162 yellow-flowered plants and 18 pink-flowered plants.

(b) (i) Calculate the frequency of yellow-flowered plants in this population. (1)

(ii) Calculate the frequency of the allele **a**. (2)

(iii) Calculate the expected number of homozygous pink-flowered plants (**AA**). (1)

(c) The habitat was surrounded by fields that are free from primroses. Explain why this increases the confidence that can be placed in the calculated genotype frequencies. (2)

(d) Pink-flowered primroses are frequently cultivated in gardens. Explain why the Hardy–Weinberg equations should **not** be used to calculate the allele frequencies for **A** and **a** in garden populations. (2)

Examiner tip

Frequency measures how relatively common items are. Having a frequency of 1.0 means there is only one type present.

Hint

Use the Hardy–Weinberg equations and use the chart to check your answers.

Answers to quick check questions

The quick check questions are designed to test your understanding and recall of material in the spreads. You will find many of the answers quite easily. However, some are more difficult and require you to put several ideas together. Answers to quick check questions are given here.

Unit 1 – Communication, homeostasis and energy

Module 1 – Communication and homeostasis

Communication and homeostasis (page 2)

1 To monitor changes, respond adaptively and coordinate activities
2 Electrical signals in neurones; hormones.
3 See the definition on page 2.
4 See the definition on page 2.
5 Negative feedback restores the original state. Positive feedback makes changes from the original state even greater.
6 They can result in unstable, often extreme states.
7 See the flow diagram on page 3.
8 See the flow diagram on page 3.
9 See the flow diagram on page 3.

Nerves (page 4)

1 They convert one form of energy into another.
2 See the table on page 4.
3 See one of the examples on page 4.
4 A sensory neurone carries impulses from a receptor to the central nervous system; a motor neurone carries them from the central nervous system to an effector.
5 See the bullet points on page 5.
6 Both are long cells; have cell bodies and axons; carry nerve impulses; may be myelinated. A sensory neurone has a long dendron; a motor neurone has many dendrites.
7 Myelin speeds up the transmission of a nerve impulse.

Transmission of a nerve impulse (page 6)

1 Actively transporting K^+ ions into and Na^+ ions out of the cell; the organic anions in the cell; the slight loss of K^+ ions from the cell.
2 See the bullet points for action potential on page 6.
3 Depolarisation makes the p.d. across the membrane *less* negative (and eventually positive) inside with respect to outside. Hyperpolarisation makes the p.d. across the membrane *more* negative inside with respect to outside.
4 The membrane becomes depolarised ahead of the action potential.
5 Impulses 'jump' from one gap in the myelin to the next.
6 Neurones from particular receptors pass to particular parts of the brain.
7 By the number of cells responding to the stimulus and the frequency of action potentials from those cells.

Synapses (page 8)

1 A synapse is a functional contact between neurones; a neurotransmitter is a chemical that passes a signal from one neurone to another.
2 See the diagram on page 8.
3 See the list on page 8.
4 See the bullet points on page 9.
5 The neurotransmitter hyperpolarises the post-synaptic membrane, so that the depolarisation is not sufficient to reach the firing threshold.
6 *Summation* is the additive effect of impulses at a synapse. In *facilitation* one action potential leaves the postsynaptic membrane more responsive to the next.

Hormones (page 10)

1 Exocrine glands have ducts; endocrine glands do not.
2 A molecule released into the blood that acts as a chemical messenger.
3 See the bullet points on page 10.
4 Hormones are first messengers. Second messengers are chemicals released in a cell as a result of a hormone binding to the cell's membrane.
5 See the table on page 11.
6 They are small islands of endocrine tissue surrounded by exocrine tissue.
7 The exocrine tissue secretes digestive enzymes; the endocrine tissue secretes insulin and glucagon.

Regulation of blood glucose (page 12)

1

Receptors	Effectors
α and β cells; receptors in plasma membranes of liver, muscle and fat cells	Liver, muscle and fat cells

2 See the flow diagram on page 12.
3 Insulin is released by exocytosis, triggered by depolarisation of the cell' s membrane. This requires ATP-sensitive K^+ ion channels.
4 Insulin-dependent diabetes is Type 1; insulin-independent diabetes is Type 2 (see page 13).
5 Five advantages are listed on page 13.
6 The porous capsules allow hormone molecules out, but not antibody molecules in.

Module 2 – Excretion

Excretion (page 14)

1 To remove potentially harmful waste products.
2 See the flow diagram on page 14.
3 The upper 'cut' surface of the diagram of a lobule is a transverse section.
4 Hepatic artery and hepatic portal vein → sinusoids → hepatic vein.
5 See the diagram on page 15.
6 See the diagram on page 15.
7 Renal artery; afferent arteriole; glomerulus; efferent arteriole; capillaries in cortex and medulla; venule; renal vein.

Kidney function (page 16)

1 See the labels A to E of the diagram on page 16.
2 Podocytes support the ultrafiltration membrane.
3 To reclaim substances the body cannot afford to lose.

4 Osmosis. The filtrate has a higher water potential than the blood from which it was filtered.
5 The activity of the loop of Henle decreases the water potential of tissue fluid in the medulla. Hence water leaves the collecting duct by osmosis when ADH is present.
6 Receptors in the hypothalamus detect changes in the water content of blood. When the water content falls they stimulate the posterior pituitary gland to secrete ADH, which allows water uptake from the filtrate in the kidney.
7 Dialysis and kidney transplant.
8 Untreated oedema can be fatal because the tissue fluid that is not being removed from the tissues contains toxins (such as urea) that will inhibit enzyme action and kill cells; the accumulation of tissue fluid means that the volume of circulating blood is reduced, making it hard to pump.

Module 3 – Photosynthesis

Photosynthesis (page 18)

1 See the definition on page 18.
2 See the flow diagram on page 18.
3 See the equation on page 18.
4 Light-dependent reactions on the lamellae; light-independent reactions in the stroma.
5 The large surface area allows for more photosynthetic pigments, more light absorption and more electron transfer.
6 See the diagram on page 19.

Photosynthesis: light-dependent reactions (page 20)

1 ATP, reduced NADP and oxygen.
2 *Absorption spectrum*: absorbance of light by a pigment; *action spectrum*: effectiveness of different wavelengths of light for photosynthesis.
3 c. 540 nm.
4 They absorb a range of wavelengths of light and pass energy to the primary pigment.
5 See the table on page 21.
6 In cyclic photophosphorylation electrons return to the primary pigment from which they were emitted. In non-cyclic photophosphorylation electrons pass elsewhere.

Photosynthesis: light-independent reactions (page 22)

1 The molecule (RuBP) that combines with carbon dioxide. It 'fixes' carbon dioxide.
2 RuBP + $CO_2 \rightarrow$ unstable 6-carbon intermediate molecule.
3 ATP supplies energy for the reduction of GP to triose phosphate and for the regeneration of RuBP. Reduced NAD supplies hydrogen needed for the reduction of GP to triose phosphate.
4 Triose phosphate forms RuBP, hexose and other carbohydrates, lipids and amino acids.
5 It is used to regenerate RuBP.
6 10 molecules (30 carbons) to 6 RuBP; 2 molecules (6 carbons) to 1 hexose or to fatty acids or amino acids.

Limiting factors (page 24)

1 The variable that limits the rate of a process.
2 Any three from; light intensity; temperature; water supply; stomatal opening; availability of chlorophyll, carriers and enzymes.
3 Light intensity is the limiting factor.
4 Some factor other than light intensity is limiting, such as concentration of carbon dioxide.
5 See the second graph on page 24. The added curve should plateau in the same way *below* the others.
6 Light intensity is the limiting factor.
7 See the diagram and text on page 25. (See also page 38.)

Module 4 – Respiration

Respiration: ATP (page 26)

1 See the definition on page 26
2 Examples of work are listed on page 26.
3 It is efficient because using a single compound makes it easy to control and coordinate different energy-requiring processes.
4 *Energy currency*: the molecule that is the immediate source of energy for reactions in the cell – ATP. *Energy storage*: a large, compact, more or less insoluble energy-rich molecule – amylose, glycogen or triglyceride.
5 (ii)
6 See the equation on page 27.
7 Muscle contraction; active transport; synthesis of proteins such as myosin or actin (see page 86).

Respiration: glycolysis (page 28)

1 In the cytoplasm (cytosol).
2 A 6-carbon sugar.
3 To make glucose more reactive so that its chemical potential energy can be trapped more efficiently.
4 The coenzyme NAD is a hydrogen acceptor.
5

Used per glucose	Produced per glucose	Net gain per glucose
−2 ATP	+4 ATP	+2 ATP

6 Two molecules each of: pyruvate, reduced NAD and ATP.
7 It passes to the link reaction in a mitochondrion.

Aerobic respiration (page 30)

1 See the equation on page 30.
2 It transfers an acetyl (2-carbon) group from pyruvate into the Krebs cycle.
3 (i) and (ii) mitochondrial matrix; .(iii) cristae of mitochondrion.
4 Oxygen is the final electron acceptor and is reduced to water.
5 The production of ATP by a flow of protons through ATP synthase.
6 See the diagram and explanation on page 31.

Anaerobic respiration and respiratory substrates (page 32)

1 (i) Yeast; (ii) mammalian muscle.
2 See the flow diagram on page 32.
3 The disadvantages (and one advantage) are listed on page 32.
4 Two roles are listed on page 33.
5 See the definition on page 33.
6 See the flow diagram on page 33.
7 Lipids have more hydrogens per molecule than have carbohydrates.

Unit 2 – Control, genomes and environment

Module 1 – Cellular control and variation

Cellular control: protein synthesis (page 42)

1 Transcription of DNA makes mRNA; translation of mRNA makes polypeptide/protein.
2 $4 \times 4 \times 4$ [i.e. 4^3] = 64.
3 DNA provides a sequence of bases that code for the *primary structure* of the polypeptide. RNA polymerase matches free RNA nucleotides against template strand of DNA to make messenger RNA (transcription). tRNA molecules carry amino acids to ribosomes. Each tRNA is specific to a particular amino acid. At the ribosome, anticodons (on tRNA) pair with codons (on mRNA). Amino acids are joined by peptide bonds (translation).

4 tRNA anticodons: CCU, AGC, UUC.
5 There are no anticodons for the three stop codons. Each of these codons signifies the end of a sequence of bases coding for a polypeptide.

Mutation and the phenotype (page 44)

1 An unpredictable change to the coding of a gene.
2 Translation will finish at the stop triplet, giving a shortened, non-functional, product.
3 Addition or deletion of one nucleotide alters all subsequent triplets (frame shift). Some may now code for different amino acids but some may be altered to stop signals.
4 The altered triplet may still code for the same amino acid.
5 Sickle-cell anaemia: oxygen shortage because distorted red blood cells block capillaries and are destroyed.

Regulation of gene activity (page 46)

1 Transcription is production of mRNA from DNA. RNA polymerase is needed to join RNA nucleotides into mRNA.
2 It avoids wasting energy making an unused enzyme.
3 Glucose reduces the concentration of cAMP, inactivating a transcription protein.
4 In the absence of cAMP the transcription control protein is the wrong shape to be active.
5 A sequence of DNA found in many genes that control body plans.
6 Two examples are given on page 47.

Meiosis, genetics and gene control (page 48)

1 Meiosis I has crossing-over and separates homologous chromosomes; meiosis II separates sister chromatids.
2 Both separate sister chromatids and produce two daughter cells.
3 Portions of non-sister chromatids exchange places, giving new combinations of alleles.
4 2^n different combinations; n = 4, so 2^4 [i.e. 2 × 2 × 2 × 2] = 16 genetically different spermatozoa.
5

	Mitosis	Meiosis
Number of division cycles	1	2
Number of daughter cells	2	4
Number of chromosomes per nucleus in daughter cells	16	8

Genetics (page 50)

1, **2** and **3** See definitions on page 50.
4

		Gametes from red parent
		C^R
Gametes from pink parent	C^R	C^RC^R red flowers
	C^W	C^RC^W pink flowers

Equal numbers of red and pink flowers are expected, i.e. a ratio of 1 red: 1 pink.

Dominance and sex linkage (page 52)

1 It is the only allele of that gene present, since the Y is so small.
2 X^hX^h
3

Parental phenotypes	Haemophiliac carrier	
	Man	Woman
Parental genotypes	X^hY	X^HX^h
Genotypes of gametes	(X^h) or (Y)	(X^H) or (X^h)

Genotypes and phenotypes of offspring:

		Gametes from woman	
		(X^H)	(X^h)
Gametes from man	(X^h)	X^HX^h Carrier female	X^hX^h Haemophiliac female
	(Y)	X^HY Normal male	X^hY Haemophiliac male

A ratio of phenotypes 1:1:1:1 is expected.
4 The interaction between different genes in which one gene masks the expression of another.
5 aaBB and aaBb
6 aaBB, aaBb, aabb

χ^2 test (page 54)

1 4 classes of data. $\eta = (4 - 1) = 3$.
2 Not due to chance. Some other factor is responsible for the results.
3 $\chi^2 = 2.67$. $\eta = (2 - 1) = 1$. The probability of $\chi^2 = 2.67$ is greater than 0.1. The results are due to chance.
4 $\chi^2 = 25$. $\eta = (4 - 1) = 3$. The probability of $\chi^2 = 25$ is < 0.001. The results are not due to chance. The expectation of a 1:1:1:1 ratio is incorrect. Another explanation is needed.

Phenotypic variation (page 56)

1 Continuous variation does not have the clearly distinguishable classes of phenotypes shown by discontinuous variation.
2 A number of genes are involved, having additive effects. Different alleles of a particular gene have small effects.
3 Phenotypic variation is a combination of genetic and environmental variation.
4 V_E is the phenotypic variation caused by variation in the environment.
5 The gene is only expressed at the cooler temperatures of the extremities.
6 Those in low light intensity will be less vigorous, less green, have thinner but longer stems.
7 AA and Aa will have the same phenotype thanks to dominance.
8 Homozygous recessives = q^2 = 16% = 0.16, so $q = \sqrt{0.16} = 0.4$
So $p = 1 - 0.4 = 0.6$ and $p^2 = 0.36$ = 36% (the homozygous dominants)
Heterozygotes $2pq = 0.48$ = 48%

Genetic drift and natural selection (page 58)

1 A chance change in allele frequency in a population.
2 Stabilising selection maintains allele frequencies in a population; directional selection alters them.
3 See the definition on page 58.
4 Allopatric speciation involves geographic isolation; sympatric speciation does not.
5 Isolation allows genetic drift in a small population and selection for different adaptations from those of the main population. When enough differences have built up the population may be reproductively isolated.
6 Doubling the chromosome number of a sterile interspecific hybrid allows it to be fertile, but with a different chromosome number from each of its parent species.
7 They will separate into different species on each of the different fruit trees.

Artificial selection (page 60)

1 Selection of an organism's traits that are of use to the breeder.
2 See the table on page 60.

3 It is difficult to select for desirable alleles of many genes when choosing organisms as parents. There is often a large environmental effect on the phenotype.
4 Performance testing compares different organisms under the same conditions; progeny testing compares the offspring of, say, different males to assess the value of the male for breeding.
5 The bull does not show the trait (it is sex-limited) and so his value as a sire can only be judged by his daughters' performance.
6 The chromosomes cannot form homologous pairs during prophase I of meiosis.
7 Doubling the chromosome number gives each chromosome a homologue with which to pair in prophase I of meiosis.

Module 2 – Biotechnology and gene technologies

Cloning in plants and animals (page 62)

1 See the definition at the top of page 62.
2 Elms are propagated by means of suckers or by layering.
3 A mass of undifferentiated cells.
4 The advantages and disadvantages can be found in the table on page 62.
5 A female mammal that has been treated with hormones so that a donated embryo can be implanted in its uterus. She may be the same species as the donated embryo or a related species.
6 A nucleus from an adult tissue cell is put into an enucleate egg cell.
7 The chromosomal DNA is human and the mitochondrial DNA rabbit.

Biotechnology (page 64)

1 See the definition at the top of page 64.
2 Two of: depletion of nutrients or oxygen; accumulation of waste products.
3 Limiting factors slow the growth until the death rate equals the division rate.
4 The advantages and disadvantages are listed on page 65.
5 Batch culture and continuous culture are compared in the table on page 65.

Genomes and gene technology (page 66)

1 A ddNTP is a nucleoside triphosphate (NTP) which lacks the hydroxyl group necessary to continue extending a growing polynucleotide chain.
2 A primer forms a starting point for DNA polymerase to replicate a strand of DNA.
3 A set of fragments of DNA, differing in length by one nucleotide and each ending with a nucleotide with a different base.
4 The ability to label each of the four types of ddNTP with a different colour fluorescent dye and to detect the different colours.
5 DNA fragments are negatively charged.
6 The shorter a fragment of DNA, the further it travels in the time allowed.
7 The steps of the procedure are listed on page 67.

Gene technology (page 68)

1 and **2** See the definitions at the top of page 68
3 The ways in which a gene is obtained are listed on page 68
4 A restriction enzyme is a bacterial enzyme which cuts the DNA of invading bacteriophage viruses.
5 Blunt ends do not have the single chains of unpaired nucleotides of sticky ends.
6 Denaturation, annealing and elongation.
7 A vector delivers a gene into a cell.

Genetic engineering (page 70)

1 Reverse transcriptase uses mRNA as a template to produce cDNA. Restriction enzymes cut DNA at specific target sites. Plasmids are used as vectors.
2 Sticky ends with complementary bases form hydrogen bonds.
3 This transformation allows useful genes to be shared.
4 The Ti plasmid of the bacterium was the vector which added the wanted genes to the rice genome.
5 Gene therapy is the treatment of a genetic disorder by altering an individual's natural genome.
6 The therapies are distinguished in the diagram on page 71.
7 Ethics change because people alter their views according to their knowledge and experience.

Module 3 – Ecosystems and sustainability

Ecosystems (page 72)

1 See the table of definitions on page 72.
2 Secondary consumers eat primary consumers.
3 See the table of definitions on page 72.
4 The energy losses from food chains are listed on page 73.
5 Energy content is found by burning dry samples in oxygen in a calorimeter to measure the heat energy released.
6 Human activities affecting energy flow through ecosystems are listed on page 73.

Succession (page 74)

1 Particles provide gaps into which spores and seeds can slip; allow rhizoids and roots to penetrate; give air spaces; hold water.
2 The adaptations of pioneer plants are listed on page 74.
3 The typical features of plant species in a climax community are listed on page 74.
4 When there are too many organisms to count, and when they are distributed relatively evenly in uniform conditions.
5 Subjective scales are very quick to use.
6 **(i)** Random quadrats and identification; **(ii)** random line transects/belt transects at a right angle to the river and identification.

Decomposers, the nitrogen cycle and carrying capacity (page 76)

1 Decomposers feed on other organisms' waste or on dead organisms.
2 Protein is hydrolysed to amino acids, which are broken down to release NH_4^+. This is converted by nitrifying bacteria to NO_3^-, which is taken up by plants.
3 The roles of the bacteria can be seen in the diagram on page 76.
4 Biotic factors involve other living organisms; abiotic factors involve non-living components of the environment.
5 The effect of density-dependent factors is proportional to the size of the population.
6 Carrying capacity is defined on page 77.

Predator–prey interactions and competition (page 78)

1 The peaks in hare numbers come *before* peaks in lynx numbers.
2 Troughs in prey species lead to decreased numbers of owls. Peaks in prey numbers in autumn and early winter stabilise existing numbers of owls. (The owls do not increase in numbers then as it is not their breeding season.)
3 Intraspecific: within a species; interspecific: between species.
4 As a result of competition, two species will not occupy the same niche.
5 *Paramecium aurelia* may be a more efficient feeder than *P. caudatum*. *Lemna minor*, which floats, reduces the light reaching the submerged *L. trisulca*.
6 Activity of the N-fixing bacteria in the clover's root nodules is suppressed by presence of nitrogenous fertiliser in the soil.

Conservation and management (page 80)

1 Conservation involves active management to maintain or improve biodiversity; preservation simply protects species or habitats.
2 Conservation measures are listed in the table on page 80.
3 Sustainable management allows an area to be exploited indefinitely; unsustainable management results in damage or depletion.
4 Selective felling cuts down some trees; clear felling cuts down all the trees on a site.
5 Conservation of the islands, education and scientific research.
6 Three of: increased tourism and colonisation, introduced predatory or competitive animals, resentful islanders.
7 Wildlife attracts paying tourists, but their presence and their need for accommodation and support threatens wildlife.

Module 4 – Responding to the environment

Plant responses (page 82)

1 A sensor or receptor, a hormone and an effector.
2 A growth response in a direction determined by the stimulus.
3 Three of: auxin, gibberellic acid, cytokinin, abscisic acid, ethene.
4 The tip of a cereal coleoptile.
5 The region of cell elongation of a shoot, coleoptile or root.
6 The apical bud of a plant inhibits the growth of lateral buds.
7 The sequence of events is listed on page 83.
8 Three examples are listed on page 83.

Animal responses (page 84)

1 The central nervous system (CNS) is the brain and spinal cord. The peripheral nervous system is the nerves between the CNS and the rest of the body.
2 Sensory neurones and the motor neurones to skeletal muscles.
3 Sympathetic and parasympathetic motor neurones to glands and muscles *other* than skeletal muscles.
4 and **5** See the diagram labels on page 84.
6 Pairs of muscles, attached across a joint, which pull in opposite directions.
7 Flexor muscles bend a joint; extensor muscles straighten it.
8 Contraction of the biceps and brachialis muscles; relaxation of the triceps.

Muscle contraction (page 86)

1 A muscle fibre is a muscle cell; myofibrils are organelles made up of actin and myosin filaments.
2 The endoplasmic reticulum of a muscle cell (fibre).
3 See the diagram on page 86.
4 The sequence of events is listed on page 86.
5 ATP provides the energy to separate actin and myosin.
6 Myosin heads (which are ATP-ases) move actin filaments towards the centre of the sarcomere.
7 Creatine phosphate provides a phosphate group to convert ADP to ATP.

Muscle structure and function (page 88)

1 Release of ACh from the presynaptic membrane; diffusion of ACh; binding of ACh to receptors on the sarcolemma causing depolarization.
2 See the upper table on page 88.
3 See the lower table on page 88.
4 In the walls of tubular structures such as the gut, blood vessels and ducts.
5 Increase the pulse rate and stroke volume of the heart; cause secretion of adrenaline.
6 The effects are listed on page 89.
7 Association areas correlate sensory information to initiate appropriate responses.

Animal behaviour (page 90)

1 Innate behaviour is genetically determined; learned behaviour is modified by experience.
2 Escape reflex, kinesis, taxis.
3 The advantages are listed on page 90.
4 Habituation: an animal stops responding to a repeated stimulus; imprinting: learning to recognize a parent.
5 Classical conditioning: learning to respond to an unusual stimulus; operant conditioning: learning to perform a particular action for reward or to avoid an unpleasant experience.
6 Exploratory behaviour not directed to a particular outcome.
7 Chimpanzees using past experience of play with boxes to reach fruit.

Social behaviour in primates (page 92)

1 Behaviour associated with living in a group.
2 Basic needs are provided more effectively by living in a group.
3 Cooperation to scare predators and to hunt.
4 Cooperation with a relative increases the chances that the alleles they have in common will survive.
5 DRD4 5R has 5 repeats of a 48 bp sequence; DRD4 7R, 7 repeats.
6 Each allele codes for a receptor with a different 3D shape.
7 Examples are listed on page 93.

Answers to end-of-unit questions

Unit 1 – Communication, homeostasis and energy

1 **(a)** Basal metabolism is needed to produce ATP/release energy to keep body alive/maintain cellular activity; physical activity/exercise requires energy/ATP; (aerobic) respiration produces ATP/releases energy; blood transports oxygen/glucose needed for respiration/carbon dioxide produced in respiration; heart rate rises to increase blood flow, to transport glucose/oxygen/carbon dioxide; (3)

(b) Vasoconstriction of blood vessels/arterioles in the skin; blood flow to the skin surface/to surface capillaries restricted; heat loss to cold air/environment reduced/heat retained within the body; (2)

(c) (i) Heat gained must be lost through the skin/body surface; if skin is warmer than blood, the body will gain heat; blood transfers heat from muscles to surface (or equivalent statement); (1)

(ii) Muscle contraction produces heat; heat is a product of (increased) respiration; respiration increases during muscle contraction/as muscles work harder; (2)

(iii) At rest, the intestine, blood/artery and muscle are all just below 38 °C; this is the *core* body temperature; blood and intestine remain very close to 38 °C during the gallop; (1)

(d) Release of adrenaline from adrenal glands/adrenal medulla; adrenaline causes dilation of arterioles/blood vessels in skeletal muscles/leg muscles; increased blood flow to leg muscles brings oxygen and removes CO_2/lactate; adrenaline causes constriction of arterioles/blood vessels in the gut; so blood is diverted from gut to muscles; cardiovascular control centre stimulated by increased blood CO_2/decreased pH; cardiovascular control centre in medulla oblongata; increased frequency of impulses in sympathetic neurones/nervous system; also results in constriction of gut blood vessels; decreased frequency of impulses in the parasympathetic/vagus neurones/nerve; sino-atrial node of heart stimulated; by adrenaline in blood/noradrenaline from sympathetic nerves; heart rate increases; stroke volume/force of heart contraction increases; rise in blood pressure; detected by receptors on (carotid) artery walls; negative feedback via vagus nerve; keeps blood pressure from increasing (to dangerous levels); vagus nerve inhibits sino-atrial node; reference to dilation of skin blood vessels/arterioles; (9)

2 **(a)** The radioactivity increased; by about 1000 counts min^{-1}; (2)

(b) Axon/axon membrane became more permeable to sodium ions; (radioactive) sodium ions entered/rushed into/diffused into the axon; through open (voltage-dependent) sodium gates/channels; because of an electrochemical gradient; resulting in an action potential (and increased radioactivity); (3)

(c) (i) Radioactivity declined at a steady rate; from about 1200 counts min^{-1} to about 400 counts min^{-1}; because sodium ions were being actively transported/pumped out of the axon; by the sodium–potassium pump;

(ii) The radioactivity did not change significantly/remained at about 2700 counts min^{-1}; because no ATP was available for active transport/for the sodium–potassium pump; as the inhibitor prevented respiration; (4)

(d) (Seven hours after the start of the experiment), sodium entered the axon, in the presence of the inhibitor; (1)

(e) (i) Synapses; (1)

(ii) light microscope cannot resolve them (however great the magnification); because the wavelength of (visible) light is not short enough; (1)

(f) Calcium ions enter axon; cause synaptic vesicles to fuse with presynaptic membrane(s); ACh/acetylcholine released; by exocytosis; ACh diffuses across synaptic cleft; binds to receptor proteins/ligand gates; in the post-synaptic membrane(s); positive ions/sodium ions pass through receptors into nerve endings; this lowers/changes the resting potential; if resting potential reaches the threshold, an action potential will result; (6)

3 **(a)** 17 000 ÷ 15; ×1130 approximately; (2)

(b) (i) Blood pressure/hydrostatic pressure in the capillaries is high/higher than in other capillaries; the glomerular capillaries are surrounded by fluid/filtrate rather than cells; support prevents bursting; (2)

(ii) Gaps allow fluid/filtrate to leave the capillary; (1)

(c) Arteriole supplying the glomerulus is wider than the arteriole taking blood away; naming these as afferent and efferent arterioles; this makes pressure higher than in most capillaries; capillary walls have pores/are fenestrated; basement membrane is permeable to water and molecules smaller than the plasma proteins; plasma proteins remain in the blood; osmotic forces not great enough to overcome the (raised) hydrostatic pressure of the blood; (4)

4 **(a)** Ascending wall cells actively transport ions; this requires ATP; which is produced (by aerobic respiration) in the mitochondria; (2)

(b) Water is absorbed into the blood; the blood plasma becomes more dilute/its water potential is increased; this sensed by osmoreceptors in the hypothalamus; the posterior pituitary gland produces less ADH; the collecting duct walls become less permeable to water; (3)

(c) A gerbil would need to produce more concentrated urine; because it would need to conserve water; longer loops of Henle allow lower water potential/more concentrated tissue fluid, in the medulla; (2)

(d) They diffuse into the blood capillaries; and leave the kidney via the renal vein; (2)

5 **(a) (i)** The ornithine cycle; the liver; (2)

(ii) Urea is a chemical substance made by liver cells, urine is a solution containing many substances (including urea) (which passes out of the body); (1)

(iii) Meat contains lots of protein; excess amino acids are deaminated; (2)

(b) (Much) less toxic than ammonia; does not require as much water as ammonia to dilute it in the urine; (2)

(c) (i) From the fatty acids excreted by the bacteria; (1)

(ii) Total amount of protein in the kangaroo and the bacteria does not increase (on cellulose diet); some nitrogen is lost/excreted/bacteria do not get all the urea produced; breeding/pregnancy would require additional protein; because it involves growth/new cells/cell division; the bacteria cannot fix nitrogen/use atmospheric nitrogen; protein needed in large amounts for growth of foetus/baby; and for lactation/producing milk; (3)

6 **(a)** oxygen was produced during photosynthesis; increasing the pressure in the syringe/apparatus; (2)

(b) movement in opposite direction/towards the syringe; because the cells would absorb oxygen during respiration and the carbon dioxide produced would dissolve in the water;
or no movement of the bead; because the volume of oxygen used in respiration would be balanced by the carbon dioxide produced; (2 max.)

(c) **(i)** (the rate of movement would increase steadily/ linearly as the temperature increased; because when the cells are in bright conditions, the Calvin cycle/ carbon fixation/light-independent reactions limit photosynthesis; these enzyme-controlled processes go faster at higher temperatures; 20 °C is unlikely to cause damage/denaturation;

(ii) rate would only increase to a small/smaller extent; light dependent reactions/synthesis of ATP/ reduced NADP would be limiting; bead might start moving towards the syringe at higher temperatures; because respiration would increase/become faster than photosynthesis; (5 max.)

(d) **(i)** ^{14}C incorporated into organic compounds/combines with RuBP; converted to other substances/sugars in the Calvin cycle; chain of enzyme controlled reactions; (2 max.)

(ii) glycerate phosphate (GP); (1)

7 **(a)** **(i)** chloroplasts can be moved to places in the cell with optimum light intensity;

(ii) maximum chance for light/photons to be absorbed (by a chloroplast); large surface/volume ratio;

(iii) light passes down cell for a maximum distance; (3)

(b) temperature; humidity; water supplied to cuttings; day length; concentration of nutrients/ions/named nutrient; (2)

(c) genotype is constant/no genetic variation; all cells produced by mitosis; the plants/leaves are from one clone; (1)

(d) **(i)** both sets are below the light compensation point; respiration is faster than photosynthesis; (2)

(ii) in the cells developing at low light intensity (or converse argument), lower concentrations of enzymes; fewer/lower density of mitochondria; fewer/lower density of chloroplasts; less chlorophyll; lower concentrations of NAD/NADP/P_i/ADP; reduced rate of respiration; becomes light saturated at low light intensity; at intensities above about 800 lux; (4)

8 **(a)** **(i)** products of the light-dependent reactions; photophosphorylation; Z scheme; ref. to photosystems I and II; in grana (of chloroplasts); (3)

(ii) used again in the light-dependent reactions/recycled; (1)

(b) RuBP is constantly used and reformed/in the Calvin cycle; (1)

(c) triose phosphate is transported out of the chloroplast, condenses to form hexose/glucose phosphate; from which glucose, starch (and sucrose) are produced; (2)

9 **(a)** Fewer CO_2 molecules are able to react with RuBP; so fewer RuBP molecules are used; to produce GP, which falls in concentration as a result; RuBP concentration then falls because RuBP is produced from GP; (3)

(b) Light (energy) is required to produce ATP and reduced NADP; in the light dependent reactions/by photophosphorylation; RuBP is converted to GP, so RuBP concentration falls and GP concentration rises; ATP and reduced NADP no longer available to allow GP to be converted to RuBP; (3)

10 **(a)** **(i)** A = stroma; B = thylakoid; C = granum; (3)

(ii) P – the ribosomes; M – on the thylakoids; W – inside any of the thylakoids;
Z – on any of the granal membranes; R – in the stroma;
Y – in the stroma; (6)

(b) **(i)** the light dependent reactions reduce NADP; which then reduces the methylene blue; (2)

(ii) both are (strongly) absorbed by chlorophyll; (1)

(c) **(i)** electron microscope only usable on dead/dehydrated specimens; (1)

(ii) move to points of optimum light intensity/carbon dioxide concentration; (1)

11 **(a)** **(i)** Slow down chemical reactions/prevent enzymes/ proteins from denaturing; (1)

(ii) Keep pH stable/optimum pH; (1)

(iii) Prevent organelles taking in water by osmosis/keep water potential constant; (1)

(b) **(i)** A = nuclei; C = ribosomes; (2)

(ii) Aerobic respiration; (1)

(iii) Glycolysis takes place in the cytoplasm/cytosol/not within any organelle; (1)

(c)

Treatment given to cell fraction B	Effect on rate of oxidative phosphorylation	Suggested explanation
Increasing temperature from 5 to 15 °C	Increase	More molecular collisions/nearer to optimum temperature for mammal
Add glucose	No effect	Mitochondria do not use glucose
Add pyruvate	Increase	Reference to link reaction/pyruvate is taken in by mitochondria
Bubble oxygen through the fraction	Increase	Oxygen is the final electron acceptor/it reacts with electron and hydrogen ions to form water

(8)

12 **(a)** Glucose is converted to pyruvate; in glycolysis; reduced NAD is produced (in glycolysis); this is re-oxidised when pyruvate is converted to ethanol; and CO_2; (3)

(b) **(i)** Ethanol molecules are small/soluble /diffuse easily /do not need to be hydrolysed (before absorption); (2)

(ii) Reduces/passes electrons to other molecules; used in aerobic respiration/by mitochondria/passed to the electron transport chain; used in lipid/fat /fatty acid synthesis; (2)

(c) Drugs/surplus amino acids/haemoglobin; (1)

13 (a) (Highly) folded; cristae; large surface area; electron carrier proteins/ATP synthase distributed throughout; presence of cytochrome oxidase; phospholipid bilayer not permeable to H^+; (3)
(b) ATP is usually hydrolysed to provide energy (but here the electron transport provides energy); (1)
(c) Chemiosmosis; (1)
(d) (i) React with oxygen; to produce water; by means of cytochrome oxidase/last electron carrier; (3)
(ii) Matrix; (1)
(e) Glycolysis/link reaction/Krebs cycle; (2)
(f) (i) Chloroplast; (1)
(ii) Glycolysis/Krebs cycle; (1)
(g) Did not explain H^+/proton gradient/electron carrier proteins did not produce ATP when purified; (1)

Unit 2 – Control, genomes and environment

1 (a) (i) Provides energy (which is needed to produce the tRNA–amino acid bond and finally the peptide bond); (1)
(ii) Amino acid must be linked to tRNA to be brought to a ribosome; which moves along an mRNA molecule; the tRNA binds to the ribosome; if its anticodon is complementary to a codon on mRNA; the tRNA places the amino acid in the correct position to be added to one end of a growing polypeptide; amino acids must enter the polypeptide chain in the correct sequence; (4)
(b) (i) Rough endoplasmic reticulum; (1)
(ii) Rough ER has (many) ribosomes; which synthesise protein from the labelled amino acids; the proteins pass into the tubes/spaces/of the ER; (2)

2 (a) RNA does not replicate; it is broken down by the cell after a relatively short time/DNA is never broken down; (mutant) DNA passed on to daughter cells/new organisms; (2)
(b) (i) Not all DNA codes for the amino acid sequence of a protein; the change must be within a gene to change the protein; different triplets/codons stand for the same amino acid (so these codons are interchangeable); correct reference to a *degenerate* code; the mutation may never be transcribed if the cell has already become specialised/differentiated; (3)
(ii) The sequence of bases/nucleotides, codes for the amino acid sequence/primary structure; different codons; (1)
(iii) Ribosomes move three nucleotides along mRNA (each time an amino acid is added to the growing polypeptide); deletion or addition of three nucleotides/bases will result in a single amino acid change; deletion or addition of one nucleotide/base will result in all codons/triplets that follow being different; this is a frame shift; (3)

3 (a) Long-rayed plants breed true/produce only long-rayed offspring; no-rayed plants also breed true/produce only no-rayed offspring; short-rayed plants produce offspring of all three types; in the ratio of 1 long : 2 short : 1 no ray; (3)
(b) Suitable genetic symbols, e.g. R^n = no ray and R^l = long ray; long-ray and no-ray plants homozygous; both these produce only one genotype of gamete/eggs/ovules/pollen; short-ray plants are heterozygous; produce two genotypes of gamete; codominance of the two alleles; (5)
(c) All short-ray; (1)
(d) Cross-pollination/fertilisation of one long-ray ovule/egg by a single no-ray pollen grain; a short-ray seed getting into the wrong container; mutation (of a long-ray allele to a no-ray allele); (2)

Examiner tip

In an exam paper there would be a space allowing or requiring a genetic diagram.

4 (a) 1 round : 1 wrinkled; (1)
(b) For round grains, 25.5 μm; for wrinkled grains, 7.3 μm; (2)
(c) Move slide without looking down microscope, measure grain nearest the graticule division/pointer/centre of field; (1)
(d) Stain with iodine solution; (1)
(e) Different cells receive different amounts of sugar/carbohydrate; environment in which grains form is variable; not all cells produce the same amount of enzyme; grains may not all be same age/stage of development; (2)
(f) α glucose; (1)
(g) Boil the extract and then show that it will no longer produce starch; change the pH and demonstrate that the reaction has an optimum pH; (1)
(h) ATP; (1)
(i) Concentration/amount of glucose-1-phosphate; volume of extract used; mass of peas used to extract the enzyme; temperature; pH; (2)
(j) Structure of enzymes depends on genetic code/genes/genotype; reference to nucleotide sequence of DNA; round and wrinkled seeds have different alleles; therefore code for slightly different proteins; which synthesise starch at different rates; round allele codes for more active/efficient enzyme; (2)

5 (a) Seed production involves fertilisation; cross-fertilisation/cross-pollination/brings together alleles/DNA from different parents; gametes (for fertilisation) produced by meiosis; gives rise to genetic recombination/independent assortment of alleles; seeds more widely dispersed; suckers grow as a result of mitosis; so form a natural clone; (4)
(b) All the resulting plants are genetically identical; so will behave/grow/respond uniformly/show less variation; will be similar to the parent tree; which can be selected for its high yield; all the plants will be susceptible to same diseases; loss of genetic diversity; improved variety cannot be selected from the plants; (4)
(c) Both mammal and plant cloning starts with specialised cells; plant cells/tissues (grown in culture) can be stimulated to develop into (many) complete plants; mammalian tissues do not; mammalian embryos can be subdivided; term totipotent used for cells able to differentiate into any kind of specialised cell; hormones are used in plant cloning; ref to gibberellins/auxins/kinins; mammalian cell nucleus must be placed in egg/oocyte cytoplasm; after removal of the egg nucleus/DNA; nuclear transfer not needed for plants; plant embryos can develop *in vitro*/in agar culture; mammal embryos must be transferred to a surrogate mother/cannot develop in culture; plant cloning relatively easy, mammal very difficult; only a very small percentage of cloned mammal embryos develop normally; (7)

6 (a) Because bacterial enzymes are pH-sensitive; so bacteria have an optimum pH for growth/cell division; sensor linked to a computer/data processor; which controls inlet valves; so that either acid or alkali can be added to the medium; to reverse pH changes: this is negative feedback; (4)

(b) (i) Enzyme molecules would be carried away in the sewage effluent flow/could not be recovered/used again; they would be less stable/more likely to be denatured in solution; (2)

(ii) Advantages: larger beads have more space/larger gaps between for solution to flow past; offer less resistance to flow; are less likely to be carried away in flow; Disadvantages: larger beads have smaller surface-to-volume ratio; so trimethylamine/substrate will not pass/diffuse into the beads so quickly; (3)

(c) (i) Curve B does not rise so steeply at lower temperatures; has a lower peak; more flattened peak; does not decline as steeply at higher temperatures; (3)

(ii) Enzyme and substrate molecules move faster/have more kinetic energy at higher temperature; immobilised enzyme cannot collide so easily with substrate/trimethylamine; so rate of reaction lower; especially at lower temperatures; substrate/trimethylamine molecules diffuse faster into beads at higher temperatures; immobilised enzyme vibrates less at high temperatures; therefore more thermostable/less easily denatured; (3)

7 (a) Restriction enzyme; sticky ends; ligase; (3)

(b) The original bacteria were all free of the plasmid; so any that are resistant to A must have a plasmid; (2)

(c) Because gene 2 has been disrupted/damaged by the insertion of the new gene; (1)

(d) A promoter allows/controls the transcription/activity of a gene; the promoter of gene 2 becomes the promoter of the new DNA sequence/new gene; (2)

8 (a) *Artemisia* and rice are very dissimilar/unrelated species of plant; cannot be crossed/will not produce fertile offspring; (1)

(b) Isolate the gene from *Artemisia* cells; using restriction enzymes/reverse transcriptase and mRNA; introduce the gene into an *Agrobacterium* plasmid; using DNA ligase; infect rice cells/tissue/embryos with *Agrobacterium;* use a marker gene/herbicide/antibiotic-resistance gene; to select cells containing the plasmid; grow plants from the infected tissue; test the plants for their temperature tolerance; (5)

(c) Unnatural; may be (unexpectedly) toxic/harmful; precautionary principle; genetically engineered crop may cross with a weed; cross with the plants of organic growers; may be harmful to wildlife; produced for profit/will not benefit poor/people from less developed countries/farmers; (2)

(d) (i) Substance produced by a plant that is not essential to its healthy growth; benefits plant by resisting some environmental stress; may be defences against herbivores/pathogens/fungi/bacteria; (2)

(ii) Taxonomic group of (very) similar species, having many common features; (1)

9 (a) Combine with cell membranes; combine with/denature proteins/enzymes; (1)

(b) Woodlice, 3 : 1; springtails, 1 : 3; (2)

(c) (i) The larger number of litter animals/detritivores in Wetmoor Wood break down the litter more quickly; bacteria/microorganisms/fungi may be inhibited in Hallen Wood; (1)

(ii) The cadmium/heavy metal has adversely affected a parasite/predator of springtails; springtails are tolerant/resistant (reject: immune) to cadmium; competition between springtails and other detritivores is less; (2)

(d) (i) Plants $5.0 + 3.8 + 14.2 = 23.0 \div 3 = 7.67$ mg kg^{-1}; animals $56.5 + 119.3 + 125.7 + 41.1 + 56.6 = 399.2 \div 5 = 79.8$ mg kg^{-1}; (2)

(ii) Animals eat more than their own mass/weight of plant; reference to pyramid of biomass; animals at higher trophic level concentrate materials gained from lower levels; (2)

(iii) Different niche; feed on different things; active/develop at different times; occupy different microhabitats; live for different lengths of time, longer life gives more time to accumulate cadmium; may differ in ability to absorb/excrete cadmium; (2)

(e) Reference to quadrats/point quadrats; grid and use of random number table; suitable measure of abundance (frequency/density/percentage cover); (2)

(f) There was originally a low concentration of cadmium in the wood; concentration increased (as a result of pollution from smelter); original population sensitive to cadmium; very low frequency of genes/alleles/genotypes that give tolerance; reference to mutation giving rise to tolerant genotypes/alleles conferring tolerance; mutation random/not caused by the cadmium; tolerant genotypes have no advantage/may be disadvantageous when cadmium concentration is low; tolerant genotypes survive/grow relatively well in presence of cadmium; therefore become more frequent; reference to selection/natural selection; recombination/crossing between different tolerant genotypes, may result in greater tolerance; as cadmium concentration increases over many years, alleles producing tolerance completely replace those that result in sensitivity; (6 + 1)

10 (a) (i) So that there are (young) leaves to eat; (1)

(ii) Increasing day length; increasing light intensity; higher temperatures; (2)

(b) Tits do not find/eat all the caterpillars; not all of the caterpillar is digested/some parts lost as faeces; some energy lost in excretion; loss in respiration; movement/maintaining body temperature; (4)

(c) (i) Inverse correlation/as distance increases, chance of predation decrease; trend almost linear; this is because weasels search for nests in the places where they are most abundant; positive correlation between population of predator and prey; density dependant factor; (max. 2 if only described/explained) (3)

(ii) The maximum population/population density that a particular habitat can support; about 250; (2)

(d) It is active management of the habitat/community/ecosystem; because it is removing a limiting factor; will also allow other bird species to nest; reducing competition (for nests); (2)

11 (a) Limited space, so a small plant is more convenient/takes less room /limited nutrients available; (1)

(b) To compare with each of the experimental treatments; the control plants are treated in the same way as the experimental ones in every way except for the treatment given/to allow for confounding variables; (2)

(c) (Mean) height of plants is decreased and number of branches increased; 14 cm shorter than controls, and number of branches (almost) doubles; (2)

(d) Hormone A had no significant effect; results in rows 2 and 3 of the table similar; hormone B reversed the effect of removing the terminal buds; restored apical dominance; comparison of data from rows 1, 2 and 4; (4)

(e) The hormone treatment would not change the DNA/genes/genotype; (1)

(f) Hormone A is a gibberellin; it causes dwarf genotypes to develop normal height; hormone B is an auxin; it restores apical dominance/stimulates the development of adventitious roots; (4)

(g) See page 83 of this guide; (2)

12 (a) Substances found in synaptic vesicles; which are released into the synaptic cleft; stimulating/inhibiting the postsynaptic neurone; by binding to receptors; (3)

(b) The abnormal dopamine concentration and the illness may have a common (unknown) cause; correlation does not establish a (direct) cause; people with the same dopamine concentration may show (very) different symptoms; schizophrenia may be multifactorial/result from several risk factors; (2)

(c) Different receptor structures will bind with dopamine in (slightly) different ways/to different extent; producing different levels of stimulation/inhibition; making it more or less likely that nerve pathways/behaviours/reflexes will be activated; (2)

(d) Both molecules have similar shape and bind to same receptor; both have amine group/(benzene) ring; (2)

13 (a) (i) Mating in rats is innate, while in monkeys it is learned; the behaviour of the rat is inherited, but in the monkey it develops as a result of experiences during development; the interaction between male and female may be more complex in monkeys than between rats; (2)

(ii) Provide warning of predation/allow more extensive searches for new food sources/older members of group remember food sources or dangers/mutual grooming/conflicts avoided by social hierarchy; (1)

(b) Imprinting; ensures that a sparrow spends time courting only other sparrows/birds similar to its parent; (2)

(c) (i) The bees learned to associate two stimuli; the unconditioned (unconditional) stimulus of sucrose with the conditioned (conditional) stimulus of blue colour; they did not associate their previous (random) action with a reward; (2)

(ii) Have several tiles/backgrounds of different colours; but having a similar shade/equally light or dark; feed bees only on the blue backgrounds; observe if they then visit empty dishes on only the blue backgrounds; repeat but feed (a fresh batch of) bees on the red background; (4)

14 (a) (i) Y; (1)

(ii) The scale would be reversed, with 0 on the left and 1 on the right. (1)

(b) (i) $162 \div (18 + 162) = 0.9$; (1)

(ii) square root of the frequency of the yellow-flowered phenotype; 0.95; (2)

(iii) 0.0025/zero; (1)

(c) The Hardy–Weinberg equations assume that there is no migration into the population; pollen/seeds of primroses unlikely to come in from other sites; (2)

(d) The Hardy–Weinberg equations assume no selective advantage to either allele; the gardener may prefer/artificially select one colour of primrose; a garden is unlikely to be reproductively isolated; plants/seeds may be imported/planted from outside/from other populations; (2)

Index